the Vitality of Faith

the Vitality of Faith

Murdo Ewen Macdonald

ABINGDON PRESS
New York • *Nashville*

THE VITALITY OF FAITH

Copyright © MCMLV by Murdo Ewen Macdonald

Library of Congress Catalog Card Number: 56-8741

SET UP, PRINTED, AND BOUND BY THE
PARTHENON PRESS, AT NASHVILLE,
TENNESSEE, UNITED STATES OF AMERICA

TO

My Father and Mother

PREFACE

THE PREACHER OF THE GOSPEL, THOUGH HE DECLARES TRUTHS that are eternal—the same yesterday, today, and forever—must also address himself to the age in which he lives. Religion is always conditioned by the contemporary culture through which it speaks.

In the first few centuries of the Christian Era, preachers used the thought categories and idiom which enabled them best to communicate the message. Politics and law placed a high premium on oratory; the pulpit followed suit. In the Middle Ages logic became predominant, so preaching had a strong rational bias. After the Reformation, when the authority of Scripture had displaced that of the pope, sermons became mosaics of biblical quotations. In the eighteenth and nineteenth centuries, when the rounded period, the purple passage, and studied rhetoric after Gibbon and Macaulay prevailed, preaching tended to conform.

But within the last fifty years our cultural pattern has not surprisingly imposed a further change. The spoken word tends to be less flowery, more pointed and direct, sometimes even staccato. In such an age the polished sermon, weighed down with literary allusions and learned abstractions, is unlikely to win or hold an audience, and this must stand as my apology for the sermons in this book.

It goes without saying that I have made no effort to achieve literary effect. As I do not preach from manuscript, the direct manner of speech has conditioned the method of preparation and presentation. Although they have been edited to some extent to cast them in a form more suitable for publication, this is the way in which they have been delivered to my congregation in St. George's West Church, Edinburgh.

Preaching: "30 min. to raise the dead in" Ruskin

What, then, is preaching? "It is thirty minutes to raise the dead in," said Ruskin, reminding us of the sense of urgency that should inspire it. To preach is not essentially to insinuate, to suggest, or to propose, but to proclaim. While preachers must strive to increase their knowledge, to assimilate the best in contemporary culture, and to adapt themselves as far as possible to their age, they are commissioned, not to come to easy terms with the world, but to offer it the terms of God's decisive deliverance. This is what the New Testament calls the kerugma. If these sermons may claim to have one unifying thought, this is it. Whether in fact it emerges with any clarity, the reader must judge for himself.

Of the following chapters, numbers three and four have appeared in the *British Weekly,* and number six in *The Sacramental Table,* published by James Clarke & Company, Ltd. They are reproduced here through the kindness of the publishers and the permission of the editors.

MURDO EWEN MACDONALD

CONTENTS

9

CHAPTER ONE

The Signs of the Times

*"O ye hypocrites, ye can discern
the face of the sky; but can ye not
discern the signs of the times?"*
—MATT. 16:3

EVERYONE KNOWS THE STORY OF RIP VAN WINKLE, THE man who went up to the mountains, slept twenty years, and came back to find a new generation goggling at his enormous nails and long white beard. But most people miss what is the most significant feature of the whole story—the sign above the inn in the little town by the Hudson River. When Rip departed, the sign had a picture painted on it of George III of Britain; when he returned, the picture was of George Washington, first president of the United States. The most significant part of the familiar tale is not that Rip van Winkle slept twenty years, but that he slept through a revolution.

This was what puzzled Christ with regard to his own generation. In his coming the greatest revolution of all time had taken place, and men slept through it. It is a contradiction which he sharpens for us by means of a brilliant little parable:

When it is evening, ye say, It will be fair weather: for the sky is red. And in the morning, It will be foul weather today for the sky is red and lowring. O ye hypocrites, ye can discern the face of the sky; but can ye not discern the signs of the times?

In those primitive days weather forecasting in observing the face of nature was the most advanced form of science known to man. Is it straining truth too much then to claim

11

that what Jesus meant was this: In the realm of scientific observation man is extraordinarily intelligent, but in the realms
of spiritual awareness he can be astonishingly obtuse.

One illustration should suffice. Bertrand Russell has become
a household name; an eminent logician--which undergraduate
has not been exposed to his *Principles of Logic?*—an illustrious
mathematician who has made original contributions, a brilliant philosopher of great output (*A History of Western Philosophy* and *Human Knowledge: Its Scope and Limits,* to mention only two volumes) ; a master of style, with a flair for pithy,
lucid, powerful English prose, you would expect him to speak
with intelligence on every subject. Well, listen to his summing
up in his last book, *New Hopes for a Changing World.*

Man now needs for his salvation only one thing: to open his
heart to joy, and leave fear to gibber through the glimmering darkness of a forgotten past. He must lift up his eyes and say: "No, I am
not a miserable sinner; I am a being who, by a long and arduous
road, have discovered how to . . . master natural obstacles, how to
live in freedom and joy, at peace with myself and therefore with all
mankind.

There speaks a man with the mental powers of a giant and
the spiritual faculties of a child.

Jesus seems to imply that men who are endowed with sufficient
intelligence to observe the sequence of natural phenomena
are also endowed with spiritual insight keen enough to enable
them to discern values that are invisible. As believing Christians,
committed to fight the hydra-headed monster we call evil, we
should be sufficiently alert to know not only the fundamental
doctrines of the faith but also to sense the trends and currents
of our time.

We live in a world of maddening complexity, in an age
of doubt and upheaval, in a generation which has lost its
bearings, and as we scan the face of the contemporary skies,
what signs of hope do we discern? What evidence, if any, do
we glean that the position is not irretrievable and that the

Christian faith, far from crumbling, is beginning to assert itself with its old dynamic vigor?

THE COLLAPSE OF THE SUBSTITUTES

In Germany during the war some of us were astonished to discover that ersatz jam looked, at first sight, like real jam. It was just as red, just as sweet, and every bit as palatable. But when you exposed it to the air for a few days it lost its flavor, a great deal of its color, and became dry and gritty in the mouth. Something similar has happened to the substitutes man worshiped in place of Christianity. At first they were sweet and colorful and deceptively real, but they have long since lost their flavor. They have become bitter, dust and ashes in the mouth.

Fifty years ago men worshiped "Progress." It was the god of the Victorians, and the scientists and philosophers chanted in happy chorus that it was inevitable. No thinker of the first rank talks in that vein now for fear of ridicule. He knows that the belief in automatic progress is buried in the melancholy archives of the past.

Forty years ago men hailed science as the new religion. Like so many other religions, she could boast of miracles—electricity, the automobile, the airplane, and wireless telegraphy, to mention only a few. Then came two devastating wars, and men saw that science, far from being a saviour, was a satanic enemy in the hands of cruel, irresponsible children. "We came to see," said Reinhold Niebuhr, "that it could sharpen the fangs of ferocity as well as alleviate the pangs of pain." Anybody who believes, these days, that by itself science can save this broken, battered world must be incredibly old-fashioned in his thinking.

Up to a decade or so ago, Humanism was the most fashionable of all religions. Julian Huxley defines it as "human control by human effort in accordance with human ideals." His creed could be summed up in four simple words—"I believe in man." We should discount what the sheltered philosophers say and listen rather to the verdict of the surviving inmates of Belsen

Camp or the salt mines of Siberia. The truth is it was man, not the abstract entities we call science, economics, or politics, that plunged this world into hell and destroyed human happiness.

The Change in the Psychological Atmosphere

A doctor, well on in his sixties, recently told me that when he was a university student it was difficult to believe. Religion was on the defensive, under attack from every conceivable angle. There was the aftermath of Darwin's theory of evolution, the corroding effects of higher criticism, the different outlook fostered by the new astronomy. Atheism was fashionable, and among many became the mark of intellectual emancipation. These by themselves did not disprove religion, but they created a psychological atmosphere which militated against it.

Today all that is changed. Atheism is out of fashion and has ceased to be an ego-booster. On the contrary, many of the leaders in various realms of thought gain, rather than lose, prestige by confessing their faith in public. Aldous Huxley, though still far from orthodox, has confessed in one of his novels to a "change of heart"—he now believes in a world unseen. Eddington and Jeans in physics and the late C. E. M. Joad in philosophy and T. S. Eliot in poetry have made their confession of faith. Public figures like Montgomery and Eisenhower at the height of their military glory were not ashamed to let the world know what they inwardly believed. I am led to understand that in our universities there is shown, at the moment, a greater sympathy to religion than at any time since World War I.

I believe that C. E. M. Joad's conversion was genuine enough, but I also believe that it was accelerated by this change in the psychological climate. Joad had his finger on the pulse of the age. He was like a seismograph recording the latest tremors of public opinion, and when at length he turned to Christianity, he knew the atmosphere was favorable. The fashion has changed, the tide has turned, the spirit of the age is more sympathetic. Though personal faith is still difficult, the intellectual climate is much more congenial.

THE PRESSURE OF MORAL NEED

The great philosopher Immanuel Kant claimed that we were all born with "a categorical imperative" in our moral make-up. It is a difficult, cumbersome phrase which he broke down into the simpler words, "I ought, therefore I can." That is true, but in experience the feeling more often than not is, "I ought, but for the life of me I can't." That, friends, is the core of the moral dilemma—the chasm that yawns between the "I ought" and the "I can" side of personality. It is the awareness of such a chasm which human efforts have failed to bridge that drives men in the end to religion. This was the inner contradiction that drove despairing men in the past to the feet of Christ. Giants like Paul, Augustine, Pascal, and Tolstoy. It is still faith's greatest ally, and because it is the ever present reality the church of Christ will never go down.

When William James was at the very pinnacle of his power, he was struck down by a curious malady, a nervous upset which took the form of insomnia and deep depression. He himself was a doctor—he had been in fact a professor of medicine at Harvard University—but he could not cure himself. He sought out the best medical treatment the world could give him in Vienna, Berlin, Paris, London, and Edinburgh, but he returned to America, worse, if anything, than ever. There were times when he thought of suicide. At last in distress he went to an uneducated faith healer. To William James, professor of medicine, world-famous psychologist, and eminent philosopher, this was anathema. His professional code and training and status all cried out against it, but desperate personal need silenced the academic protests. The simple man put his hands on the professor's head and prayed. William James felt a mysterious energy thrilling and tingling through his body, followed by a sense of peace, and he knew he was cured.

This is the Christian secret weapon. The culture of the age may be quite antagonistic to the Christian claim, and far from congenial. Men may pose acute theological doubts and speak contemptuously of the organized church, but the desperate moral dilemma—the gulf between the "I ought" and the "I can" sides of their nature—will drive them willy-nilly to seek a cure. Man today, tormented within and defeated by circumstances without, would cry from the depths with Peter, "Lord, to whom shall we go? thou hast the words of eternal life."

The Pressure of a Contemporaneous Christ

There was a day long ago in Jerusalem when cunning men thought they had rid themselves of this troublesome Christ. Herod and Caiaphas signed the death certificate and saw the dead body sealed up in a tomb with a massive stone on top, and they went home rubbing their hands with glee. Jesus had been silenced—they would live in peace. Little did they know that within a few days the most revolutionary movement in history would be, in a way, launched by men who claimed to have seen him and hailed him in his risen power.

And men are still as obtuse, still as blind, still as incapable of discerning the signs of the time. They do not see that Christ is still a force to be reckoned with, by far the most revolutionary power in the world today. They shut their eyes to the revival of religion in America, the packed churches in eastern Germany, in Hungary, and Czechoslovakia, to contemporary events which are all familiar to us, a desire for a deeper international understanding, the emergence of a clearer social conscience, the phenomena of a united Church of South India and the ecumenical movement throughout the world. They fail to sense in these the pressure of an active, living, contemporaneous Christ.

At the Yalta Conference, Roosevelt and Churchill brought combined pressure to bear upon Stalin to relax his strangle hold on the Christian religion in Russia, and apparently he promised to do so. Think of the significance of such a discussion. Roosevelt, the president of the most powerful nation on earth, Churchill speaking for the greatest commonwealth of history, between them pleading the cause of a persecuted peasant 2000 years after His death by crucifixion, amid the exigencies of a global war, and modern history's most ruthless dictator acceding to their request. Do you think a dead Christ could do that?

Jesus is not a spent force. He is the Eternal Contemporary. He is forming and fashioning history before our eyes today. We scan the face of the contemporary skies and we discern his purpose, unbroken and undefeated. We sense we are on the threshold of great events. We put our ear to the ground and we hear Christ's conquering hosts on the march. So, pledging our allegiance afresh we sing:

> O Lord and Master of us all:
> Whate'er our name or sign,
> We own Thy sway, we hear Thy call,
> We test our lives by Thine.

CHAPTER TWO

A Gospel to Be Proud Of

"I am not ashamed of the gospel of Christ." —Rom. 1:16

SHORTLY AFTER THE COMPLETION OF THE WAR IN 1945, A German pastor was taken through Belsen Camp. Halfway through he stopped and burst into tears. "I am ashamed of my country," he cried. It is sad to feel ashamed of the country of your blood and birth but it is even more tragic to be ashamed of your religion, the religion in which you were suckled and reared; for this is the very foundation of the civilization which has conditioned your culture and determined your best ideals. Yet the majority of people are ashamed of their faith; not, mind you, in any vociferous or violently atheistic manner, but they do let you know—and that quite unmistakably—that Christianity is not the thing that affords them the greatest excitement in the world. It is all very well as a side interest among many others, they imply, but as a way of life, as a solution to the world's crying needs, as an all-embracing, dominating ideal under which you must subjugate all else—No! that is not for them.

Even good Christians, who deep within their hearts believe that the only hope of the world lies in the Christian religion, are so influenced by the prevailing synthetic paganism of the day that, when they do summon up enough courage to mention the name of Christ in private conversation, or in public, they do so in such a strained manner that all present are embarrassed.

To some extent we are all ashamed of the gospel of Christ. Why? What has happened to the old thrill and rapture that

18

Christianity used to stir in the human heart? Where is the former sense of awe and wonder which once gripped men, because to them—unworthy sinners that they were—had been entrusted the glorious gospel of God himself? Has the overwhelming sense of gratitude, which breathed through every page of the New Testament, vanished entirely?

Paul faced a grim and antagonistic world. He confronted a Jewish national church, entrenched in privilege and power and steeped in complacency. He went to Rome, the citadel of ruthless imperialism and rampant cynicism. He went to Athens, the center of world culture and clever philosophies, and for them all he had only one message, and that message was, "I am not ashamed of the gospel of Christ: for it is the power of God unto salvation to every one that believeth."

Is this gospel of Jesus obsolete? In this age of upheaval and desolating doubt, does it speak to us with any degree of meaning or relevance? Or is it hopelessly incongruous in this technological world, apparently at the mercy of the diabolical instruments of man's own ingenuity? Some of us do not feel ashamed of it even in this day.

A GOSPEL OF SANITY

At the very core and center of human nature there is an element of radical insanity which drives men to actions that, judged from any intelligent and rational standpoint, are plainly mad! We ourselves have known intelligent men and women who, of their own free will, became victims of dope and drink, knowing only too well what the frightful consequences would be. And you and I sometimes have, of our own volition, behaved in such a manner that any impartial judgment would brand us as crazy fools. The world in which we live seems to be infected with the same radical insanity. It has the many grim lessons of history written in letters of blood before its very eyes. Surely it does not need a further demonstration to convince it that modern war is devilish and destructive and, in the end, suicidal. Yet responsible, and to all appearances intelligent, men leading powerful nations often behave in a manner which reminds us of irresponsible children lighting matches in the middle of a pile of dynamite. If Mars is inhabited, I can imagine the government of that remote planet sending a foreign correspondent to investigate here on Earth. He surveys the dust and debris of our bombed cities: he jots down the millions of war casualties, the thousands of the maimed and mutilated: he peeps behind our iron curtains, watches with interest our sinister atomic piles and our feverish rearmament races, and hurries back to address a startled and surprised Martian Assembly: "Gentlemen! I have been forced to the conclusion the inhabitants of Earth are mad. They show marvelous technical ingenuity. They have invented radar, jet propulsion, penicillin, and have split the atom, but their favorite pastime is blowing each other to bits."

It is in such a crazy context that we dare preach Christ's gospel of sanity. The world into which Jesus came, said, "Hate your enemies, wreak vengeance on those who have wronged you, and don't be overmerciful, as it is a sign of weakness." Christ with his gospel tore that insane philosophy to ribbons, and the world has not yet got used to the revolutionary words, "Love your enemies, . . . do good to them that hate you, and pray for them which despitefully use you, and persecute you." In Fuchow, in China, there are three graves side by side. Two of them are the graves of two sister missionaries who were both murdered. When their widowed mother learned the news in Australia she was sixty-two. Immediately she sold her home and property. She went to the place where her two daughters had been murdered, learned the Chinese language, set up a school, and toiled there for twenty years. At the age of eighty-two she died and was buried beside her daughters. The gospel of Christ is the gospel of sanity.

A Gospel of Simplicity

There is a universal tendency to shake our heads in dismay at our own generation and brand it as the worst ever. We dramatize and romanticize the past, and nostalgically talk of the "good old days." But making all possible allowances for this human eccentricity, I think it is beyond argument that our own age is undoubtedly the most complex age in all history. There is the economic and political complexity, seen focused in the crises of our own day which seem to defy the reasoned solutions of skillful and experienced men. There is the cultural complexity so evident in our midst, when the old standards of value and taste in music, art, and literature seem to be in the melting pot. And there is, worst of all, the religious complexity: old landmarks of the faith are in danger of being obliterated, and the rigid moral standards, which gave strength and sturdiness to our grandparents, are now dissolving before our eyes. Pessimistic voices, not without a note of nervousness, tell us the Christian creed is untenable, and others more jubilant assure us it is dead, beyond the faintest hope of recovery. Meanwhile many men and women, who do not wish to live lives of lust or license, are looking for a reasonable set of values which will hold life together. When they see their old beliefs dissolving in the corroding acids of modern cynicism, when they watch the old foundations yielding under their feet, they search desperately for some sure standing ground. Here again Jesus, with his usual staggering simplicity, goes to the heart of the matter. To a world sunk in confusion and blank despair, he says: "Seek ye first the kingdom of God, and his righteousness; and all these things shall be added unto you"—among them international peace and world brotherhood. And to men and women defeated by their sins, and experiencing frustration and disillusionment, he says, "Put God at the center of your lives, and a unified will, moral victory, and peace of heart will be thrown up as accidental by-products" as it were.

Amid the baffling perplexities of a crowded and complex life, the decisions of Jesus were always simple. In the agony of Gethsemane he was heard to say, "Thy will be done." And when the shadow of the cross loomed over him, he set his face to go to Jerusalem. In the labyrinths of life, however complex and complicated, there is always a way out, and it is the way of the Cross. Captain Oates, crippled by frostbite and physically exhausted, was jeopardizing his companions' chances of reaching safety. They refused to leave him. It was a maddeningly complex situation, which he solved by walking out into the raging blizzard. The way of the Cross is like that—difficult and simple. Jesus confronts us in this bewildering age of ours, and offers us, not a clever philosophy, or a complicated theology, or an abstruse dialectical theory. He asks us to rise and follow him. His is a gospel of simplicity.

A Gospel of Security

A new conscience has emerged in the modern world which clamors for security. All political parties, however divergent in their ideologies and aims, offer the masses a measure of social security. During the war I met a brilliant Oxford graduate who smiled at me pityingly and said, "Your Christianity is finished, and your God is tottering unsteadily on his crumbling, theological pedestals. He has had his day. The new god, the god of the future, which every enlightened and progressive citizen will serve and worship, is the God of Social Security." We would agree that a measure of security is surely a legitimate demand on the part of ordinary people who suspect they are mere puppets driven by blind, impersonal, economic forces beyond their ken or control. But to make it the be-all and end-all of existence, and to erect it as the gilded idol of the masses, is surely impracticable, naïve, and dangerous. Social security, however well planned, can never dispel the radical sense of religious insecurity clutching at the heart of modern man—an insecurity which stems from doubting the fact of God, the basic rationality of the universe, and the essential meaning of life itself.

Our rude forefathers may have been ignorant and superstitious, but amid many accidents and calamities they felt at home in the world because they believed it was God's world and that underneath were the everlasting arms. But modern man, when he suspects God is an illusion, providence mere fancy, and the Christian creed a product of myth and make-believe, experiences a desolating sense of loneliness and isolation in a universe that cares nothing for his hopes and fears. Matthew Arnold, who also felt the assault of the ultimate doubt, describes this feeling in his poem "Dover Beach":

> . . . for the world, which seems
> To lie before us like a land of dreams,
> So various, so beautiful, so new,
> Hath really neither joy, nor love, nor light,
> Nor certitude, nor peace, nor help for pain;
> And we are here as on a darkling plain
> Swept with confused alarms of struggle and flight,
> Where ignorant armies clash by night.

Without a firm belief in God's fatherhood, man will always feel nervous, restless, and insecure in this world. When Sir James Barrie delivered his famous speech "Courage" to the students of St. Andrews University, he read to them the letter Captain Scott wrote before dying in a tent near the South Pole:

If this letter reaches you, Bill and I will have gone out together. We are very near it now, and I should like you to know how splendid he was at the end—everlastingly cheerful and ready to sacrifice himself for others. . . . His eyes have a comfortable blue look of hope, and his mind is peaceful with the satisfaction of his faith in regarding himself as part of the great scheme of the Almighty.

Bill Wilson died in a raging blizzard in indescribably uncomfortable circumstances, and yet his eyes had a "comfortable blue look of hope" in them. This alone is real security.

So far from being ashamed of our gospel, we offer it proudly to our age; to a mad and distracted world we offer a gospel of *sanity*. To a world stunned by its own maddening complexities, we offer a gospel of *simplicity*. To a world that is shot through and through with fear and uncertainty we offer a gospel of *security*. Finally, when we confront the last enemy, we shall not cringe, but cry, "O death, where is thy sting? O grave, where is thy victory?"

CHAPTER THREE

Can Human Nature Be Changed?

*"If any man be in Christ, he is a
new creature."* —II Cor. 5:17

"YOU CANNOT CHANGE HUMAN NATURE." SO MEN HAVE SAID
from time immemorial, and today there are millions of people
who would heartily endorse that view. Take any cross section of
society you like—a group of workers from one of our factories, a
collection of undergraduates from one of our universities, a
gathering of the country's most eminent scientists—confront
them with the challenge of the times and the desperate need
for a new order to replace the shambles of this shattered age,
and before long a disillusioned voice will interrupt your dis-
cussions: "Yes, it sounds all right, but there is only one snag—
you cannot change human nature."

Is it the case? This is certainly the most important question
we can ask ourselves in the New Year. No matter how brilliant
our plans, how profound and penetrating our vision, how conse-
crated our lives, or well directed our energies, will they all one
day topple over and be smashed and broken because we have
left out of our reckoning this unknown, incalculable quantity
—human nature?

This is the question, above all others, we must endeavor to
answer before we dissipate our energies, seeking to change a
world ultimately at the mercy of what Jung calls the "raging
libido" of the lower instincts. "You cannot change human nature": Is that true or is it a lie? Is this widely held belief a
dangerous myth which has blinded men's eyes, paralyzed them
at the nerve centers of their being, and held them in chains

for centuries? To answer this most cogent of all contemporary questions, we shall call upon three representatives of modern culture to give their answer.

THE ANSWER OF THE PESSIMIST

"You cannot change human nature." That, says the pessimist, is a basic truth that needs no proof. All you have to do is to glance at the blood-spattered pages of history. For what is history but a sordid, squalid record of man's inhumanity to man, his insatiable greed, his incurable pugnacity, his devilish sadism, and his suicidal urge to destroy civilization itself?

To come down from the realms of vague abstraction to that of concrete facts, think of the mass murders ordered by Ivan the Terrible, of the trail of blood and loot and misery following in the wake of Napoleon Bonaparte, of the savagery rampant in Scotland at the time of John Knox. And if you plead that all that belongs to the bad old days, think of the two global wars within living memory which killed and mutilated countless millions. That is how man has always behaved and that is how he will continue to behave until the last syllable of recorded time.

But, you protest, isn't the pessimist exaggerating and indulging in melodrama? Isn't he painting the picture with too broad a brush? Is he not forgetting the gradual conquest of barbarism and the rise and advance of culture? We are told that the Grand Inquisitor in the Netherlands burned heretics in specially prepared slow ovens to the greater glory of God! That could not happen today. Cromwell, a man of principle, devout, austere, sometimes humane, took the town of Drogheda and put all the male inhabitants to the sword, believing this to be the will of God. That could not happen in the twentieth century.

You blind, sentimental fool! What happened to you during the last two decades or so? Did you not know that things infinitely more savage than that took place in Europe—Europe with its ancient universities, its soaring Gothic cathedrals, its world-famed culture, its sacrosanct traditions? Did you see the emaciated bodies of Belsen Camp and the pile of white glistening bones lying in every corner of it? Did you hear of it? Did you hear of the Nazi torture laboratories, of their extermination centers, where some four million Jews were gassed and thousands of political prisoners were liquidated? Human nature changed? The pessimist answers his own rhetorical question: Yes, changed for the worse if history is any criterion; barbarism, naked and unashamed, stalks the roads of Europe, and any intelligent observer, struck by the stupid futility of it all, can only conclude with the arch cynic, Omar Khayyàm:

> Into this Universe, and *Why* not knowing
> Nor *Whence,* like Water willy-nilly flowing;
>   And out of it, as Wind along the Waste,
> I know not *Whither,* willy-nilly blowing.

THE ANSWER OF THE OPTIMIST

The first thing we must do is to put things in their proper perspective. Considering this planet has existed for millions of years and human history can be measured within a tiny span, we have not done too badly. Before the march of progress and the advent of knowledge some of our stubbornest prejudices and many of our most entrenched superstitions have entirely disappeared. We have jet propulsion, penicillin, pain-killing drugs, and science is pointing to possibilities hitherto undreamed of. Even if we feel far from secure, we should congratulate ourselves in having traveled a considerable distance from the primeval slime where our remote ancestors wallowed.

It is true that there are regrettable lapses. The turning back of the clock by the Nazis, the obscurantism of the Bolshevists, the gullibility of the masses; but these should not unduly disappoint us. The wise, shunning the extreme of pessimism, will continue to believe that man, unaided by the gods, under his own steam, propelled by his own ingenuity, will work out his own salvation and establish visibly upon this earth, the perfected order of his own dreams.

This gospel of optimism was part and parcel of Victorian

culture, and found measurable expression in the words of its popular scientists, philosophers, and poets. Herbert Spencer was speaking for his age when he said "progress is inevitable." Tennyson saw nothing amiss in dreaming of a time within human reach when

> . . . the war-drum throbb'd no longer,
>   and the battle-flags were furl'd
> In the Parliament of man, the Federation of the world.

This dominant philosophy, buttressed by the positive achievements of science, seeped into every conceivable compartment of life. Religion was by no means immune. All over the country congregations rose to their feet and lustily sang the confident words of the hymn:

> These things shall be: A loftier race
>   Than e'er the world hath known shall rise
> With flame of freedom in their souls
>   And light of knowledge in their eyes.

Perfection was just around the corner.

This naïve, optimistic philosophy of life has been destroyed by the facts of modern history. Men's holiest dreams have been sabotaged by this incalculable element—human nature. If there is a way out of our dilemma we must look in another direction.

THE ANSWER OF THE REALIST

There is a Gaelic proverb which says, "The twist that is in the old stick is not easily straightened," and the whole of history bears out its truth. The radical twist in human nature, the evil endemic in our constitution, the fundamental split in personality are facts which defy the slick solutions and the facile philosophies of our day.

There are commentators who assure us that all we need to

solve our crises is a novel idiom of expression, a new orientation of our energies, a keener awareness of social urges and demands, a greater sensitiveness to current intellectual trends. I submit that, in view of the gravity of the situation, that amounts to downright blasphemy.

It is at this point that the Christian gospel speaks with its own compelling cogency. It claims that what we need is neither pleasant persuasions nor clever remedies, but a revolution in the realm of personality—a revolution of such sweeping dimensions that only words like these can do adequate justice to it. "If any man be in Christ, he is a new creature: old things are passed away; behold, all things are become new."

And here, thank God, we are dealing not with conjectures or ideals or vague probabilities but with indissoluble facts which have changed the course of human events.

Paul himself is by far the best example. His life, coupled with his epic achievements, is a fact which no amount of clever debunking can ever explain away. A burning fanatic, dominated by one narrow, concentrated purpose; a savage persecutor, whose hands were red with the blood of the first Christian martyr; a man who hated Christ and his Church—Paul was all of these. He spent all his splendid energies to erase Christ's influence from the page of history. Yet in midcourse he was checked and changed beyond all recognition. The man who stoned Stephen was the man who did most to build the Christian church and make her the cornerstone of Western civilization. The apostle of hate became the apostle of love, the author of the greatest lyrical outburst of Christian love in all literature.

Such startling facts are as common today as they were in the first century of the Caesars. There is the case of C. S. Lewis, who is beyond question the most popular Christian apologist of our times. He made his first impact as an Oxford don, a typical twentieth-century agnostic. Clever, cynical, contemptuous of puritanism, he, like so many of his contemporaries, regarded skepticism as an infallible mark of emancipation.

Then he became conscious of Christ, not as an interesting historical myth, but as a living reality, and his whole life underwent a cataclysmic change. The brilliance once used to confound Christian orthodoxy is now employed to commend the Christian faith to doubting intellectuals. The iconoclast has been changed into the convincing and constructive apologist. "If any man be in Christ, he is a new creature: old things are passed away; behold all things are become new."

CHAPTER FOUR

The Meaning of Conversion

> *"Except ye be converted, and be-*
> *come as little children, ye shall not*
> *enter into the kingdom of heaven."*
> —MATT. 18:3

THERE IS A SENSE, OF COURSE, IN WHICH EVERYBODY BELIEVES in conversion. Victor Gollancz, in his autobiography, talks of his conversion to pacifism, and Arthur Koestler, in his, of his conversion to Western democracy. We would all agree that this is a legitimate use of the word, but we who profess Christ, though we use the same word, mean by it something very different. The Christian is converted not to an ideology or a political philosophy or a system of ideas, but to Jesus Christ. Christian conversion involves a direct, dynamic confrontation between two persons, the sinner on the one hand and the Saviour on the other.

So far, so good. Why then do most decent people feel strangely embarrassed when the word "conversion" is mentioned, and associate it with the kind of person who is not only eccentric but rather objectionable as well? Many reasons can be advanced for this attitude. There is the tendency among certain evangelicals to equate conversion with catastrophic change, and we are apt to be suspicious because our own experience does not coincide with theirs.

Again, there is the common belief that conversion is something bizarre and odd and on the whole unnecessary. If a child is reared in a Christian family, in a Christian community, in a Christian church, we are inclined to argue he will grow up in the knowledge of God quite naturally and need none of your decisive encounters. But we are coming to see that this popular assumption within our churches is dangerously complacent, not to say naïve. The facts of the case are driving us to agree with William James when he says, "[The crisis of] self-surrender has been and always must be regarded as the vital turning-point of the religious life."

I believe that the most important question facing the Church at the moment is not this or that particular method of evangelism but the larger problem of what precisely we mean by Christian conversion. This sermon is an attempt to answer the question as frankly and as fully as it is possible within the compass of a few pages. When puzzled people say, "What do you mean by conversion?"—I point to four ways in which conversion expresses itself.

An Emotional Revolution

Matthew Arnold's famous definition of religion as "morality touched by emotion" is defective, but at least he does see that emotion plays a prominent part. Those who think that they can be genuinely religious without experiencing any emotion whatsoever are not only contradicting the New Testament, they are flying in the face of reality itself.

We would all agree that aesthetic emotion is not only a fact but a very important one. Who is not moved by watching a setting sun lingering in the western sky, or gazing at one of the great masterpieces of art, or looking at a superb example of medieval architecture? And what sensitive man is there who has not been moved to his depths by Keats or Wordsworth, Beethoven or Bach? How then can we expect a man to contemplate his Creator and his Redeemer without emotion? Any religion that bypasses the emotions has nothing to offer to the tortured spirit of our times.

While I was in Australia I was shown round the Eildon Dam, the largest hydroelectric scheme in the British Empire. Australia's primary problem is water. In some places there is too much, in others too little. In winter the high grounds are flooded and in summer, on the plains, there are severe droughts. All these big water-power schemes are planned to trap the water before it goes to waste, and with it to irrigate and fertilize good land some thousands of miles away. So, up in the highlands of Victoria, they were in process of constructing a dam bigger than Sydney Harbour, and by means of a gigantic powerhouse and enormous pipes they propose to direct the water wherever it is most needed. It is, in a way, a picture of what happens to our emotions under the impact of conversion. Where previously there was no purpose, things are now co-ordinated and controlled, and, harnessed to one dominating purpose, are made to flow in one direction. Direction, not dissipation, is the keyword. Without this drastic reorganization in the realm of the emotions there can be no real change.

There is another way in which conversion expresses itself:

A MORAL REVOLUTION

In other words, the will as well as the emotions are engaged in the operation we call conversion. If it is real at all, it must result in a radical change in the realm of character. Sometimes it happens in a flash; more often it comes at the end of a long, desperate, agonizing struggle. Indeed, we could safely say that even the most catastrophic conversions are marked by a painful preliminary struggle, a prolonged crisis. This is exactly what Frederic Myers, in his poem on Paul, means by the verses:

> Let no man think that sudden in a minute
> All is accomplished and the work is done.
> Though with thine earliest dawn thou should'st
> begin it,
> Scarce were it ended in thy setting sun.
>
> Oh, the regret, the struggle and the failing,
> Oh, the days desolate and useless years,
> Vows in the night, so fierce and unavailing,
> Stings of my shame and passion of my tears.

One of the best and most vivid descriptions of this moral revolution I have ever read I came across in a recent book—*Candlelight in Avalon*. This story is of an intelligent, cultured man who is on the verge of a nervous breakdown. From the first he realizes that the deep-seated malaise is not physical or mental but spiritual, that the real answer lies not in a restful holiday and a nightly ration of sleeping pills but in the radical transformation of all his values. He knows that he has to be adjusted not merely to himself and to his work but to his God. In the long, painful struggle he is helped by a Roman Catholic priest, an ardent Anglo-Catholic, a nonconformist lawyer, and in the end he reaches the conviction that Christ is not only God incarnate, but also the Saviour of his soul. Whether it comes slowly, or suddenly, or in a combination of both, the distinguishing mark is always the same—"If any man be in Christ, he is a new creature: old things are passed away, . . . all things are become new."

The third way in which conversion expresses itself is:

An Intellectual Revolution

So far, I suppose all the evangelical schools of thought—be they Salvation Army, Plymouth Brethren, Kentucky Baptist, or Free Presbyterian—would agree with me. They insist on a revolution in the realms of the emotions and the will, but this is not far enough; we must go further—we demand a revolution in the realm of the intellect.

There are certain evangelicals who despise reason, scorn scholarship, and are forever clamoring for what they call the "simple" gospel. I fail to see how they can get around Christ's own command to love the Lord our God, not only with our hearts and souls and strength but also with our minds.

And here history is most decisively on our side. Men who have experienced a dramatic conversion have been foremost in blazing an intellectual revolution. Paul is the supreme example. In the experience of the Damascus Road he discovered not merely an integrating principle but also an intellectual passion.

It is no accident that he was the first to try his hand at Christian theology. Augustine's life in a way runs along parallel lines. In the orchard at noonday he was intellectually released as well as being morally transformed. He saw through the falsity of the Manichaean philosophy of which he had been such an eloquent exponent. He became the intellectual giant whose mind still towers above us. The same is true of men like Luther and Calvin. Their minds, as well as their hearts and wills, were pierced through by the light from another world. They cut loose from contemporary orthodoxy and medieval prejudice and gave their age a new and vigorous system of theology.

And the challenge today is no less pressing. In an age when science dominates, when passionate ideologies rage, when the philosophy of collective materialism sways the minds of millions, we Christians must offer to the world a better and more relevant picture of reality. Our salvation is supernatural, but it is never irrational. It commends itself not only to our wills and hearts but also to our reason. It generates energy not only on the level of the emotions but also in that of the intellect. A great many prejudices have to be demolished by a realistic theology before Christ will come into his own.

And the fourth way in which conversion expresses itself is:

A SOCIAL REVOLUTION

Though men like John Wesley and Thomas Chalmers had a sharp social conscience, the disturbing thing is that the majority of ordinary evangelical Christians separated personal religion from social concern, and today we are reaping the nemesis of that short-sighted divorce. Communism, sweeping like a plague across the earth, is a judgment—a grim reminder that we cannot keep the religion of Christ in a private, personal watertight compartment—that it has revolutionary repercussions that reach out to the whole of society.

I have just written a review of David Livingstone's travels, based on his own diaries. It is a thrilling book in which the real stature of Livingstone finally emerges. And the thing that

35

gripped me most was not the courage, the grim determination, the phenomenal capacity for observation of this truly heroic man, but the fact that, as far back as 1850, he saw more clearly than any of his contemporaries that evangelical religion must issue in a social revolution. Other explorers like Bruce and Park had recorded and regretted the scandal of the slave trade; only Livingstone fought it with a burning sense of righteousness. When he told Cambridge University that commerce and Christianity must go hand in hand, he did not mean that Christian missions were the prelude to economic exploitation: this man, burning with a passion to blaze Christ's name abroad, was also burning with a passion for social justice.

And every Christian who has experienced the redemptive power of Christ must be like-minded. It is sheer blasphemy claiming to be personally converted and not seeing that another revolution is needed in society. The hydrogen bomb will not stop Communism, nor the Atlantic Alliance; only a more dynamic and demanding revolution is personality within and society without. The safety and ultimate security of our world depends on a redefinition of Christian conversion in principle and a living out of it in practice.

CHAPTER FIVE

Is Religion Big Enough to Save the World?

"And he shall send them a saviour, and a great one, and he shall deliver them." —Isa. 19:20

AMONG MANY SHARP AND STARTLING CONTRADICTIONS FACING us in our day, one of the most baffling is the failure of Christianity to make an impact on our world. Men are wistfully seeking for salvation. Like Isaiah's contemporaries they are peering through the shadows, desperately looking for a Saviour. This confused, yet clamorous demand for a deliverer is emphasized by some of the most acute and penetrating modern prophets. Aldous Huxley assures us that man today is asking himself the ancient, haunting question: "What must I do to be saved?" C. E. M. Joad says: "That the seeds of spiritual revival are germinating in the minds of the people of this country, I for one do not doubt." And Jung asserts that the problems of every man and woman over forty are basically religious.

This is one side of the contradiction. Here is the other. We have a Saviour who claims to satisfy this deep and desperate need of the human heart. From the very beginning he has been hailed as the strong deliverer, the mighty one who was to come and ransom his people and save them from their sins. This is what the Church has always believed, what she still vehemently and passionately proclaims; yet nothing momentous ever seems to happen. The vast majority look to politics, science, and education to solve their dilemmas, and more often than not turn away in despair and disillusionment. <u>The Galilean appears to have lost his spell</u>.

I believe that this contradiction can be explained by the fact that the gospel we have preached was not sufficiently bold and demanding, that the Christ we have proclaimed was not big enough, that the Saviour we have professed lacked the stature to commend him to men's spirits or to win their unswerving allegiance. In these critical days, when men's hearts are failing them and they are waiting for some miracle to happen from the side of God, they can only be saved by a Saviour who is great enough to meet their deepest needs. <u>Such a Saviour must make a cosmic appeal and yet touch the individual heart</u>.

1.) *I believe that the Christ we offer must be big enough to cope with the gigantic dilemmas of our world*. One of the most challenging features of our age is <u>the emergence of a world consciousness</u>. This is a comparatively new development. When Mussolini invaded Ethiopia, a member of Parliament remarked that the entire Abyssinian Empire was not worth the life of one boy from Nottingham. When Czechoslovakia was surrendered to Hitler, a prime minister had the temerity to suggest that we should not worry unduly over a small nation on the other side of Europe.

This isolationist mentality has largely disappeared. Korea is a small nation on the other side of the earth, but no one now denies that what happens there is of crucial concern to us and to our children. Today every problem is thrown into silhouette against a world background, and unless we, in like manner, widen our Christian perspective, we have no chance of appealing to the modern mind.

For centuries we have been <u>content to preach a pygmy and parochial Christ</u> who, under the pressure of big events, was forced to evacuate province after province of human experience. <u>He was made to abdicate his sovereign rights in the realm of politics, economics, science, and education, and to occupy only the limited sphere of men's private emotions.</u> We failed to see that unless Christ was the Saviour of the material as well as the spiritual, of the mind as well as the soul, he was no Saviour at all. We thought that the world's problems could be solved by a little political maneuvering, a dash of scientific enlightenment, some slight economic manipulation. We fail to see what Paul saw so clearly and vividly, that we are not up against mere flesh and blood but against principalities and powers, against spiritual wickedness in high places. We lost the New Testament insight which saw a cosmic Christ whose spirit was woven into the very stuff and essence of the universe, who, in the words of John, created all things, without whom "was not anything made that was made."

In failing to preach Christ in his cosmic totality we failed to keep step with the dynamic forces of history, and our Saviour was shorn of his stature and robbed of his power. A pygmy and private Christ is the most devastating weapon in the well-stocked armory of evil. The Nazis did not persecute religion in any crude and obvious way. They forced their pastors to preach a purely private Christ, and therefore robbed Germany of a Saviour. Stalin does not persecute religion in Russia in the barbarous manner of Nero and Domitian. Under the guise of tolerance, he permits the people a dwarfed and stunted Christ who has no power to save: a Christ who is pushed off the main thoroughfares into the narrow side-lanes of life; a Christ who is banished from the realm of men's crucial concerns—politics, education, and culture—and confined to man's inner, hidden feelings. If there is to be any hope for a world in the grip of dark and demonic forces, we must again sound the full evangel—a conquering, cosmic Christ striding forward in the greatness of his strength, mighty to save.

2.) *I believe also that the Saviour we offer must be big enough to cope with the gigantic dilemma of human doubt.* One of the most basic doctrines of the Christian religion is the doctrine of providence. It bids us believe that God's hand is on the helm, that he is the principle of rationality behind the chaos and the disorder of our world: the moral governor of our universe, the loving Father, whose everlasting arms are underneath us, bearing us up.

But can anyone blame us if we sometimes rebel and flatly refuse to believe it? Studdert-Kennedy tells how, in the first World War, he knew one old Frenchwoman who, contemptuous of the advancing Germany army, stubbornly tended her two cows and refused to move. They begged her to go, but she would not listen to reason. One day, after a fearful artillery barrage, they came across her, sitting in the middle of the road; her cows were blown to smithereens, there was blood on her clothes and face, and she was crying, "The good God is dead."

Look at the appalling human situation, at the catastrophic collapse of all civilized pretense; look at evil men deliberately planning to plunge the world into an unparalleled orgy of destruction, and can you really believe that God is active behind the scenes? In a powerful novel by Virgil Gheorghiu, the hero tells his friend of the book he is in the process of writing. "What will it be called?" the friend asks. "*The Twenty-fifth Hour,*" he answered, "the hour when mankind is beyond salvation; when it is too late even for the coming of a Messiah. It is not the last hour. It is the one past the last hour. It is Western civilization at this very moment. It is NOW."

It is while we tremble before what John Keats called "the giant agony of the world," that our hearts cry out for a Saviour big enough to turn our darkness into light and our crippling doubts into glorious, radiant certainty. It is the unanimous testimony of all the saints, the considered verdict of countless numbers of men and women through the ages, that Christ comes to them in his risen power and says as he said to Thomas of old: "Reach hither thy finger, and behold my hands; and reach hither thy hand, and thrust it into my side: and be not faithless, but believing."

This is the experience of Frank Morison, the agnostic lawyer, who set out to disprove the story of the Resurrection by gathering all the scraps of evidence and thereby showing it was all an ill-founded myth. He found, as many others have found before and since, that as he concentrated on him, Christ was no longer a ghost flitting among the tombs of the dead years, no longer a pale abstraction, a remote figure shrouded in the mists of antiquity. He jumped right out of the pages of history to meet him, a risen Saviour in whose presence he could only cry, "My Lord and my God." It takes a Christ who climbed Calvary, and who cried in anguish, "My God, my God, why hast thou forsaken me?" and thereafter rose from the dead, to carry conviction to our trembling hearts and to dispel the mists of doubt from our souls forever.

And I believe that the Saviour we offer must be big enough to cope with the gigantic dilemma of personality. If it is necessary to preach a cosmic Christ, who redeems the whole of creation, whose Spirit permeates all things, it is even more necessary to preach a personal Saviour who turns defeat to victory and saves us from our sin. We may talk eloquently of a fresh orientation of the gospel, a shedding of our narrow parochialism, a widening of our limited horizons, a magnifying of our Saviour; but if the personal note is absent, we have no message for a sad and despairing world. For here, if the trumpet give an uncertain sound, who will prepare for battle! John was able to grasp Christ's cosmic significance only because he had known Christ personally and had experienced his risen power. Paul was able to preach Christ as the Saviour of the world—of Jew and Gentile, of Greek and barbarian, male and female—because he could first say, "He saved me and caused me to pass from death unto life."

This is where the modernist Christ has been found wanting. He was not sufficiently dynamic to cope with the dilemmas of life. He was too weak for the twist of evil in human personality. He was too impotent to cope with the hard core of egotism deeply entwined in our nature. The world today offers many saviours, but they are too small to save to the uttermost—too shallow to renew the will and re-create the heart, too synthetic

41

to restore unto us the lost meaning of life and faith in the ultimate purpose of things.

We are discovering to our cost that there are no short cuts to salvation. To be saved we must be reconciled with God in the deeper levels of personality. The easy-term remedies of the substitute faiths are defied by the solid fact of sin deeply embedded in our make-up. The vicious circle of egotism cannot be broken by any human agency, but only by a Saviour stronger and mightier than man himself. In the light of this knowledge, therefore, we offer to the world the Christ of God, for "as many as received him, to them gave he power to become the sons of God, even to them that believe on his name."

CHAPTER SIX

Defeating the Brute Facts of Life

*"That I may know him, and the
power of his resurrection."*
—PHIL. 3:10

THE BIG THREE SAT IN CONCLAVE AROUND THE CONFERENCE table. Churchill, encouraged by the congenial atmosphere, suggested that a greater measure of religious tolerance in Russia would please the Pope, who naturally exercised great influence in countries predominantly Roman Catholic. Stalin pulled at his mustache, smiled sardonically, and said, "Mr. Prime Minister, how many divisions did you say the Pope has?" The story may be quite apocryphal, but it brings home forcibly the part naked power plays in modern politics. Our world is ruled not necessarily by the best and wisest countries but by those with the greatest man power and industrial capacity. Britain, proud of her long tradition and wide experience in international affairs, is forced to take a back seat and let an American general command her army and a foreign admiral control her navy, because today power speaks—and not only speaks, but has the last word.

This respect for power is by no means a modern trait. It has always had a strange fascination for the human mind, and a church that does not possess it in some form or other is not likely to make much of an impact upon our world. The early Church lacked our social prestige, our financial resources, our intellectual caliber, but she possessed in abundance what we so sadly lack—spiritual power. Her members had seen a dead Christ hauled down from a cross and laid to rest in a deep

43

tomb. They had seen a heavy stone rolled in place on top and a strong guard mounted to watch over it—yet the incredible miracle had happened. The stone had been rolled away. The tomb was empty, and Jesus had risen and revealed himself to his old companions, not as a pale, wraithlike ghost but in all his old dynamism and power. The effect upon them was startling and catastrophic. From skulking in fear behind locked doors, they marched out into a hostile world against overwhelming odds—haunted, gripped, and utterly dominated by the power of the Resurrection. This was the power that balked the plans of evil men, thwarted the forces of darkness, and raised Jesus from the dead. Therefore nothing could prevail against it. Paul, when he spoke of the power of Christ's resurrection, was not philosophizing or using figurative language. He was thinking of its effect upon himself and others as they fought in the arena of life and wrestled with the brute facts of their mortal existence.

The Brute Fact of Collective Evil

The challenging thing about the forces that hounded Christ to Calvary and crucified him there, was that they were so well organized. There was a sinister solidarity about the evil that coldbloodedly planned his destruction. Bad men, who normally would cut each other's throats, sank their differences and banded together in one unholy alliance to accomplish their nefarious ends. And as Paul marched out with the gospel of the Resurrection, he was under no illusion. Over against him stood something not vague and indefinable lurking in the shadows but the brute fact of collective evil, solidly organized and backed by the limitless might of the far-flung Roman Empire. Yet he was in no doubt as to the final issue. Collective evil, solidly entrenched and supported by named force, might appear formidable and frightening; but ultimately it was no match for the power that raised Jesus from the dead. Today the brute fact of collective evil is even more menacing than it was then.

Backed by colossal might, and inspired by demonic cunning, it has set out to conquer the world. Its agents and saboteurs are spread over all the earth, surreptitiously undermining the foundations of Western culture and Christian civilization. The militant forces of materialism are on the march and are sweeping everything from their path. They have swept over Russia and China—and, making a bold bid for the whole of Asia, are thundering at the gates of Africa, while Europe and America are in a state of frenzy and high fever. It looks as if this is an irresistible tide which nothing human can stop. Like a gigantic tidal wave it rolls omnipotently on, crashing over our pathetic, puny defenses and grinding to smithereens the dreams and hopes of the centuries. No wonder Aldous Huxley, in that gruesome book *Ape and Essence*, prophesies a time when man will have ceased to believe in the principle of good and will worship evil for its own sake, "the evil that has won the last battle in his soul." Such pessimism Paul and the first disciples would find meaningless. They had peered into the empty tomb. They had looked into the face of the risen Christ. The forces of evil might be on the march. They could do their damnedest, but they could never prevail against the power that raised Jesus from the dead.

The Brute Fact of Personal Sin

Maddeningly, Paul was aware of a principle of contradiction at the very center of his being, and he knew that a gospel which could not cope with this basic irrationality had little chance of conquering spiritual wickedness in high places. No one was better equipped to deal with this contradiction than Paul of Tarsus. Ascetic by nature, trained in the stoic virtues of self-discipline and detachment, and intensely religious, he found himself powerless in its iron grip. If penetrating insight and sheer strength of will could by themselves achieve moral victory, this genius had a flying start. But listen to the classic confession. Nowhere in literature had the problem of the divided self been more pungently and more forcibly expressed. "For the good that I would I do not: but the evil which I would not, that I do. . . . O wretched man that I am! who shall deliver me from the body of this death?" In the year A.D. 386 Augustine faced a similar inner crisis, and his description of it is nearly as striking as that of Paul's:

There was naught in me to answer thy call, "Awake, thou sleeper"; but only drawling, drowsy words, "Presently; yes presently; wait a little while," but the presently had no present, and the little while grew long. . . . I said within myself, "Come let it be done now," and as I said it I was on the point of the resolve. I all but did it, yet I did not do it.

The remarkable thing about these two accounts, the one given in the first and the other in the fourth century, is how curiously up-to-date they sound and how aptly they describe the inner convulsions of our own souls. We, too, have tried to solve our inner tensions by mortifying the flesh, by flogging the will, and by morbid introspection, but we have not been delivered. All men of insight from Plato onward have been aware of this deep-seated principle of contradiction—this lie in the soul which deceives and mocks us. This radical twist that splits human nature and runs right through the center of our being, defies all the facile cures and proffered panaceas of modern education. It clamors for the power of the Resurrection. That this power is still available and still effective no one who believes in the risen Christ must ever for one moment forget. Starr Daily, who wrote that amazing book *Release,* took to crime in his early teens. After twenty-five years he became a hardened criminal. Pronounced incurable by five psychiatrists and doctors, he was labeled as lost and damned forever by society. But one day he met Christ in prison, and what education, society, and retributive punishment failed to do, the power of the Resurrection accomplished in the twinkling of an eye. In the twentieth century, as well as in the first and fourth, the risen Christ is still "the power of God unto salvation to every one that believeth."

The Brute Fact of Death

The ancients, with their proud and robust realism, never tried to deceive themselves about death. They did not assume the pose of pseudorealism such as we find in liberal Protestant thought when it talks glibly of the immortality of the race. They did not try to explain death away from frustration to fulfillment in the way Lewis Mumford does when he argues, not too convincingly, "Continuity for us exists not in the individual soul as such but in the germ plasm and the social heritage through which we are united to all mankind and to all nature." To the ancients such sentiments would sound like sheer cowardly evasion. Death to them was a brutal fact. Across their sunlit world hung the shadow of mortality, and its grim specter haunted them and cut their laughter short; and however hard we moderns try to rationalize it, death is still the last totalitarian contradiction. In the end it all boils down to a very simple question: How do we know that this contradiction is not final? What makes us believe that it is not the irrevocable word? What reason can we give for the hope which the drenching floods of the centuries have been unable to quench? Many arguments have been adduced to carry conviction to the human heart. There is the Socratic argument which calls attention to the nature of the soul itself, indestructible in its essence and therefore proof against death. There is the argument which points to the mystery of personality and refuses to believe that human genius can be blotted out forever by a chance accident or a stray bullet; and finally there is the argument, which reminds us of the character of God himself, which claims with Tennyson that a God who is just and went to the bother of creating us will not in the end leave us in the dust or throw us out on the scrap heap. I strongly doubt if these arguments, however convincing they may sound to us, would have meant much to those who had felt the bottom knocked out of their world by the fact of Calvary. These men were eyewitnesses of the empty tomb and of the risen Christ; therefore there was no need to argue. The power that rolled away the stone and raised Jesus from the dead was theirs, and if that power was theirs, nothing, not even death, could prevail against them.

CHAPTER SEVEN

The Absurdity of Christianity

"What have I to do with thee,
Jesus, thou Son of . . . God? . . .
torment me not." —MARK 5:7

THE HEALING OF THE GADARENE MANIAC IS ONE OF THE MOST striking of Christ's miracles. He had crossed the sea in a boat to escape, if only for a brief respite, the intolerable pressure of work. No sooner had he landed on the other side than he heard the hideous laughter of a lunatic and the terrifying clanking of his chains. Emerging from the tombs, the maniac rushed toward Jesus, gesticulating wildly and crying out, "What have I to do with thee, Jesus, thou Son of . . . God? . . . torment me not!" Clearly, this man was at one and the same time strongly repelled and strangely attracted by the presence of Christ.

There are certain modern commentators who would ascribe the present chaos to what they would call the "moral lag." Technological progress has so far outstripped the advance of ethics that the world is running out of gear. Once this regrettable lapse has been rectified and our lagging morals have caught up, our most difficult problems will all be ironed out and harmony will be restored. That is one view accepted by many men of intelligence. But there is another view, less popular by far but nearer the truth in that it is more faithful to the New Testament emphasis. This view is strongly advocated, for example, by a thinker of the caliber of Nicholas Berdyaev, the Russian mystic and philosopher. It is a view that would claim that there is a demonic element at the very heart of modern culture which is the root cause of all misery. This demonism

49

resents Christianity as an alien intrusion across its own frontiers, and its attitude is perfectly expressed in the words of the text, "What have I to do with thee, Jesus, thou Son of . . . God?"

There is science, for instance, which wields such uncanny power over the contemporary mind. It may be perfectly true that the traditional divorce between science and religion is less rigid than it used to be, but it is equally true that science still tends to claim complete autonomy in its own domain. While it is busy bursting the sound barrier and making its atomic bombs, it would truculently cry, "What have I to do with thee, Jesus, thou Son of God?"

Even more obvious is the sphere of politics, which the Church in former days controlled with such iron rigidity. There was a day when the Church with a word could depose emperors, when she could at will start and stop wars and deliver ex-cathedra statements with shattering authority. That day has passed. Now she has been shunted off the main thoroughfares into a quiet corner where she is allowed to squeak out feeble protests to her heart's content. This is what happened in Germany, Russia, and China. In Britain and America we can still trounce our political leaders and appeal to the public conscience, but I wonder if the modern Machiavelli on either side of the iron curtain is in any real sense perturbed by moral principles? Is his attitude not frankly, "What have I to do with thee, Jesus, thou Son of God?"

We meet the same demonic truculence in art, which until recent times was universally regarded as the handmaiden of religion. We see this devotion enshrined in the majesty of Gothic cathedrals, their spires symbolically pointing heavenwards; in the paintings of Leonardo da Vinci and Giotto; in the poetry of Dante and Milton; in the novels of Dostoevski and Tolstoy. Beneath the various scenes and plots we can feel the pulse of a fully recognized and generally accepted Christian philosophy of life. But now you can paint and carve and write the most foul obscenity in the name of the absolute autonomy of Art. And if anyone dare protest, you can contemptuously dismiss him as puritan. Art, like the Gadarene maniac, has run amok and is crying. "What have I to do with thee, Jesus, thou Son of God?"

If modern culture, then, can claim to be autonomous, how can we ever hope to evangelize this dark and demonic world? How can we convince men that this message is both relevant and cogent?

The point of contact is to be found in the profound paradox which lies at the heart of this old story—which tells that the maniac, who so fiercely resented Christ's coming—at the same time fell down at his feet and worshiped him. It is still true that while part of us is repelled by Christ, another part of us is irresistibly drawn to him. We may disagree with him, hate him, crucify him, but we cannot ignore him. He has come to torment us.

He has come to torment us *in the realm of ideas*. There was a book written some time ago under the title *Ideas Have Legs*, which means, simply, that if an idea is powerful enough, nothing can stop it. Copernicus once conceived an idea, new, startling, and revolutionary, and no sooner was it announced than men put up barriers across its path to stop it. But it had come to torment them, and in time it stung the educated classes of Europe into acceptance.

The most masterful idea of our time appears to be that of "dialectical materialism." When Karl Marx gave it shape in the reading room of the British Museum, he could never have imagined the terrific range of its influence. How could he have foreseen on what powerful legs it would stride across the earth and with what flaming passion it would command and call forth men's allegiance. Its ultimate success will depend on whether it is more powerful than the idea Jesus launched in Galilee long ago, and which has continued to torment us ever since—the idea that the power behind the perplexing panorama of our existence is no impersonal force or blind economic urge but a Father who loves us and invests our lives with a sense of infinite significance. If this idea is false, it will inevitably die, and no amount of artificial respiration will make any difference. If it is true, nothing that evil men can do will ever stop it. This burning idea Jesus let loose upon the earth, and it will continue to torment us till we have made up our minds one way or the other.

He has come to torment us *in the realm of morals.* For centuries on end the Romans accepted the brutal spectacle of the gladiatorial shows as part and parcel of their culture. True, there were sensitive spirits like Seneca the Stoic, and Marius the Epicurean, who feebly protested; but, on the whole, the best minds in Roman society accepted it as a matter of course. Even Marcus Aurelius, austere moralist as he was, used to sit in the grandstand watching the blood and carnage with stolid impassivity. Then, one day, a Christian monk threw himself to his death in the arena as a protest against the savagery, and in this dramatic act ended the cruel practice forever. Jesus came into the world to torment us not only in the realm of ideas but also in that of morals.

Beyond question it was Jesus who gave the Western world a new social conscience—not Karl Marx. His was the spirit behind Wilberforce abolishing slavery in the British Empire, and behind Lincoln doing the same in America. His is the spirit which speaks in two entries in Lord Shaftesbury's diary which reveal the torments of a sensitive conscience. He had tried to get a bill through Parliament to protect chimney sweeps, but failed, so he writes: "Very sad and low about the loss of the Sweeps' Bill . . . but I must persevere and by God's help so I will." Thirteen years later, when the bill was actually passed, he becomes aware of the appalling conditions under which women and children worked in factories, so he writes: "The work to be done is greater than ever. . . . But surely this career has been ordained to me by God and therein I rejoice, yea, and will rejoice."

It is Jesus who makes us so desperately unhappy with our

sins. Like Burns and Byron, we may seek to cover them with cleverness or cynicism, but the more we rebel, the more clear it is that Christ has come and we can no longer sin in peace. If we could silence a restive conscience, if we would lay the inner storm of our lives to rest and experience the peace of God, we must come to terms with Christ, for he has come hither to torment us.

He has come to torment us *in the realm of personality*. We may insulate our minds against the idea Jesus let loose upon the earth. We may, like Swinburne, reject the Christian morality as too gray and grim and forbidding. We may cleverly rationalize the whole thing away and convince ourselves and others that we are immune to its appeal. But there is one thing we cannot do: see the spirit of Jesus incarnate in a human personality and remain impassive and unmoved.

Richard Hillary, in his book *The Last Enemy*, frankly confessed that he was both annoyed and challenged by Peter Peas, a fellow pilot—a professed Christian and the best man he had ever met. His one ambition was to get him alone, to attack him mercilessly and tear the fabric of his faith to pieces. His chance came as they traveled together in a railway compartment from Montrose to Edinburgh. He glared at his victim and said, "Your religion is a fake—a hereditary hangover, a useful social adjunct and no more." Peter opened his mouth, stammered out a few feeble protests, then lapsed into silence crushed by his opponent's flood of dialectics. But Hillary knew that in reality he had lost the argument, for there was one fact he could not explain—Peter's character. It and his religion were inextricably bound up together, and it played havoc with logic.

Someone recently writing of a great missionary says, "I am repelled by the man's liberal modernistic theology, but disarmed and silenced by his saintliness." As Dean Inge so succinctly puts it, "You cannot confute a person." Jesus is the great disturber of history. We can never get rid of him. He has launched an idea that will haunt us all our days. He has stabbed our conscience awake so that never again can we sin in peace. But above all, he confronts and challenges us in the lives of his saints. In the end, like the Gadarene maniac, we shall find that Jesus the Tormentor is also Jesus the Peacemaker. He alone can recover for us our lost sanity, restore to us our true selves, and make us the men and women which from the beginning we were meant to be.

CHAPTER EIGHT

The Magnetism of Christ

> *"If I be lifted up from the earth,*
> *[I] will draw all men unto me."*
> —JOHN 12:32

MATTHEW ARNOLD STOOD ONE NIGHT LOOKING AT DOVER beach glimmering in the moonlight and listened to the crunch of pebbles flung up on the strand by the incoming tide. With thoughts of the ebb and flow of human misery heavy upon him, he wrote:

> The Sea of Faith
> Was once, too, at the full, and round earth's shore
> Lay like the folds of a bright girdle furled.
> But now I only hear
> Its melancholy, long, withdrawing roar,
> Retreating, to the breath
> Of the night-wind, down the vast edges drear
> And naked shingles of the world.

So, sadly, he saw the faith of his fathers a fast dwindling legacy of the past.

In Scotland, we are told, only fifteen per cent of the population have any vital connection with the church; in England the percentage is even less; on the Continent the disease is more widespread and more serious. If this is true, how can we hope to combat the menace of materialism standing over against us, threatening to swamp our whole system of values and crush the things which to us are dearer than life itself?

The sea of faith was once at the flood. Now it is on the ebb.

That demands an explanation. Why does the prestige of the Christian church stand so low? Why is secularism making such sweeping progress? How account for the startling change in the spiritual climate of our times?

There are some who would blame the professional clergy, accusing them of too readily acquiescing in a sub-Christian culture. They have not been sufficiently alert, they have failed to anticipate the trends of modern thought, they have blunted the sharp, insistent edges of Christ's gospel and have reduced the religion of Calvary to an easygoing, amiable sentimentality. Who would dare deny that there is an uncomfortable chunk of truth in this criticism?

Some would blame the mass of ordinary laymen who stand at the strategic points where the Christian religion comes to grips with the world. They have been overcautious and over-conventional. They have shut Christ in within the narrow little circle of their own prejudices and predilections. They have barricaded their minds against the advent of new Truth and remained blind to insights which must be grasped if the Church is to survive amid the flux and flow of human history.

Others more bold would blame Christ himself. Why not face the facts honestly? they say. Jesus of Nazareth has had his day. His sun is setting fast toward the dying west and a bewildered world has started looking for a new God. There was a time when he could move the minds of men and speak straight to their hearts—but now the magic of his spell is broken and the world refuses to listen.

Perhaps there is another explanation. Perhaps we have failed to do justice to the Christ of the New Testament. Perhaps we have been guilty of proclaiming a Saviour shorn of his strength and too small and puny to cope with the gigantic dilemmas of our day. Perhaps the crucial need of our day is the rediscovery of a Christ still mighty to save and big enough to shoulder the crushing burdens of this tired and distracted world. Such a Christ looms up in the challenging words of our text, "If I be lifted up from the earth, [I] will draw all men unto me."

If Christ is to wield his ancient power; if he is to make an impact upon our age; if he is to win and draw men unto himself; we must lift him up *in the fullness of his stature*.

The Christ who threw Jersualem into a state of feverish commotion, who died and rose again and convinced his disciples of that fact beyond the faintest shadow of a doubt, was a Christ who impressed men with a terrific sense of power. They frankly confessed they could not understand him, but they had seen him in action, and to them he was King of kings and Lord of lords.

Under the impact of modern science men began to resent a Christ they could not understand, so they set out to scale him down and limit him within the bounds of logic. It became fashionable to talk of Buddha, Socrates, and Christ in that order. The Jesus-of-History school stressed his first-century limitation of outlook, the rigid Jewish thought-categories within which his mind worked. The Christ that emerged, though human and appealing, was a truncated Christ, stripped of his mystery, shorn of his strength, and powerless to save.

But it was not the devotees of science alone or the champions of modern culture who whittled down the stature of the New Testament Christ. Sincere, evangelically minded Christians, looking at things from a totally different angle, were guilty of the same thing. Not that such people ever questioned his divinity or sought to explain him in humanistic terms. They strongly and passionately declared he was God incarnate, but in presenting him to the world they restricted the scope of his salvation. They confined him to the realm of individual experience and left the rest of the world to Karl Marx and his pantheon of omnipotent dictators.

If, in the past, in Russia and Germany and in every other country, Christ had been proclaimed and preached in the fullness of his stature we would not now be cringing before the march of militant Communism. Let us by all means proclaim Christ King in the realm of individual experience, but we must go further and proclaim him King in the realms of culture and politics. If we are to appeal to the godless masses of our day, we must lift him up in the fullness of his stature.

Again, if Christ is to attract, we must lift him up *in the fullness of his saving power.*

Our minds have been so conditioned by the scientific attitude that everything in life is automatically questioned. Things which our fathers counted dear and sacred are now put under the searchlight of the intellect, ruthlessly analyzed and exposed, like so many dead specimens on a dissecting table. In a recent issue of the *British Medical Journal* there appeared an article on "The Mechanism of Conversion," by William Sargent, head of the Department of Psychological Medicine, St. Thomas' Hospital, London. In it the author draws attention to recent cases of conversion. Mrs. Haldane, wife of the noted scientist, was some time ago converted to Communism, but recently saw through the falsity, and she confesses that her reconversion experience was every bit as intense as the first. The other example he cites is Arthur Koestler, the novelist, who embraced Communism in the early thirties and enjoyed a terrific sense of inner release. During the Spanish Civil War he became aware of monstrous evil inherent in the system, and he compared his reconversion to sanity to a mystical experience. But by far the most dramatic and challenging cases are the conversions behind the iron curtain. A man is prepared for trial before the people's court, and apparently without use of torture or drugs, the whole pattern of his character undergoes a catastrophic change and in public he confesses to crimes he never committed and literally condemns himself to death. Sargent then goes on to compare these phenomena with Pavlov's experiments on animals, with examples of conversion in Wesley's day, and cases in some parts of modern America, and suggests that they are all caused by some sort of psychological excitement brought about by the use of suitably selected physical stimuli. In other words, conversion is from below, not from above. It is a displacement in the depth of the unconscious, a mere psychological upheaval, a fortuitous rearrangement of the personality pattern —not a breaking through of the divine. If this is true, we can no longer talk of Christ's saving power. He is just an unnecessary addendum tagged on to all the rest, and there is no gospel to preach.

The church of Christ must never be obscurantist. She must not stand in the way of knowledge or ever impede the forward march of truth. She must welcome all the light shed by science and psychology on the workings of the unconscious and the dilemmas of personality. But when in all humility she has made room for that, she must declare, and declare decisively, that only Christ can redeem—that there is only one name under heaven given among men whereby they can be saved. And here, mark you, she is standing on solid ground and dealing with facts hard as nails, refusing to be dissolved by the acids of modernity. She can bring out in full display the gleaming trophies of Christ's grace—Paul, Augustine, Luther, Wesley, and countless more who have passed from darkness into light and who, unable to comprehend the mystery, can only cry, like the man in the Gospels, "One thing I know, that, whereas I was blind, now I see." We must restore this note to our preaching, place it at the center of our faith, and sound it forth like a trumpet—for if we are to win defeated and despairing men we must lift Christ up in the fullness of his saving power.

Finally, if Christ is to come into his own, we must lift him up *in the fullness of his sacrificial demands.*

In Galilee long ago well-meaning friends took him aside and said: "Your gospel is too difficult. You are pitching it too high and you are not making any concession to the weakness of human nature. Tone it down a little; file away its sharp and rough edges, make it more slick and smooth and palatable, and you will win the masses." Jesus answered, "If any man will come after me, let him deny himself, take up his cross daily, and follow me." Are there not voices heard today telling us that religion is too grim and stern and forbidding? Make it more interesting and more accommodating and you'll win the young people, they assure us; and they are too blind to see that the very opposite is the case.

I sometimes believe the conversion of Constantine in A.D. 313 was one of the most disastrous facts in history. Up to then, following Christ demanded real courage. It meant persecution, social disfavor, and at times death. But by a stroke of the pen all that was changed overnight. Christianity became fashionable. There and then the rot set in and today we are reaping the nemesis. There is a current belief that compared to our fathers we are soft and spineless and decadent. That surely is a silly myth exploded by the last war and disproved by the epics of Dunkirk, the Battle of Britain, El Alamein. Whenever there was any difficult task demanding suicidal courage, volunteers tripped over one another with the ancient cry upon their lips, "Here am I, send me." Read Winston Churchill's war memoirs, and how can you believe our generation is decadent? The men who penetrated deep behind enemy lines, who dropped by parachute into hostile countries, who did the most mad and daredevil things, did it all with a gay abandon that reminds you of the courage of the early Christians. You find yourself putting down the book and saying, "If we had stuff like that in the Christian church we would take the world by storm." Yet the challenging fact remains that most of these men have little or no use for the Church. They equate the religion of Calvary with a dull, deadly, conventional respectability. In their hearts they despise it. Ah, Christ! that must be the unkindest cut of all. That must be worse than the pain of Golgotha.

And how can we win them and appeal to the spirit which will dare anything and laugh in the face of death itself? By putting the Cross back where it belongs, at the very center of faith and life. Let us confront men with the claim of Christ. Let us lift him up unashamed, in the fullness of his stature, in the fullness of his sacrificial demands—and men can no more resist him than iron filings can resist a magnet. They will rise up and follow him to the ends of the earth.

CHAPTER NINE

A Prophet's Vision

> *"Son of man, can these bones live?"* —EZEK. 37:3

THERE IS A PRESENT-DAY SCHOOL OF THOUGHT WHICH DEPRE-cates the Old Testament. Why read the annals of an old Semite tribe? Why concern ourselves with the squabbles and super-stitions of a bygone age? Why bother about platitudes and prophecies which have ceased to have any meaning for our time?

One of the best answers to these questions has been given by Herbert Butterfield, professor of modern history in the Univer-sity of Cambridge. No serious student of history, he argues, can afford to ignore the great prophets of the Bible. While believing presumably in a world beyond, they had their feet solidly planted on the earth and possessed something amount-ing to clairvoyant insight into the social and political trends of their day. History to the casual observer may appear to be a grim catalogue of disconnected, meaningless events—"a tale told by an idiot, full of sound and fury, signifying nothing." But not to those men of God. They saw, running through all the con-fusion and complexity, a divine purpose which could not be thwarted and which, in the end, turned despair to hope and death to life.

Is that not the real message of Ezekiel's famous vision of the valley of dry bones? He lived in the dark days and saw tragedy engulf the entire nation. The Assyrians had besieged and sacked Jerusalem, carrying off the prophet and the élite of society to far-off Babylon. Those left at home were of poor caliber, so stupid and shortsighted that they made the most disastrous military alliances, resulting in a second storming and sacking of Jerusalem by Nebuchadnezzar. Now desolation was complete. The last crushing blow had been delivered. There remained not a glimmer of a hope, and there was no promise of a comeback. It was then at the midnight hour of despair and utter, irretrievable ruin that the prophet's eyes were opened and he had a vision of a valley full of dry bones, where we find bleak pessimism and bounding optimism marvelously blended. In it the authentic note of a living faith clearly sounded.

The first thing that meets us here is a *gospel of realism*. Then, as now, there was no shortage of smooth sentimentalists and softheaded Utopians who kept chirping that things were not so bad as they looked, that every cloud had a silver lining, that progress in spite of disappointing setbacks was inevitable. They ascribed the eclipse of the nation to a chance turn of the wheel of fate, to a fortuitous chain of historical events, to an unfortunate political alliance—theories which the prophet lashed with a scathing contempt. "What we are witnessing," he cried, "is none of these things, but death in the region of the spirit—a drying up of the very marrow of our bones, and unless God intervenes, the last chapter has been written. This is the end."

Let us examine this vision, then, and see how it applies to our own world. Take the miracle we call Western civilization. There was a day when it simply surged with vitality, bursting out in the Renaissance and the Industrial Revolution and the rise of modern science, but now the whole grand edifice is crumbling and tottering to its ruin. So say the secular prophets of our day, the Spenglers, the Russells, the Huxleys, and the Welles. It is only a question of time till its bones are bleaching in the valley of doom and destruction.

Take the Christian church herself. There was a day, too, when she thrilled and throbbed with life and vitality, when emperors had to walk barefoot in the snow to do her homage, when she wielded enormous influence, cultural and political, when she had the destinies of nations in the hollow of her hand and designed the shape of things to come. But, alas, it looks as if that day has departed never to return. Alexander Clifford, a brilliant modern journalist, suggests in his book *Enter Citizens* that Christianity in Europe is on the way out. "God is dead," he says, "because the masses no longer believe in Him and they are clamoring for a new faith to fill the empty space in the soul." Is he painting with too broad a brush? Is he being melodramatic? Is he giving way to the current pessimism? Well, look at the church in Scotland: her gaping deficits, her dwindling congregation, her dearth of candidates for the ministry, her diminishing influence in the life of the nation; and don't you hear a voice ringing down the drafty corridors of your soul: "Son of man, can these bones live?" Is it the case that she is dying and that what we are doing is a futile effort at artificial respiration?

What is true of structures like the Church is even more true of personality itself. "The wages of sin is death," says Paul, meaning that evil strangles this will of ours. There it lies inert beyond the faintest hope of resuscitation. Brilliant men, proud of their genius, thought they could flaunt this iron law, but found it was as futile as trying to flaunt the law of gravity. It broke them in the end. Coleridge, the poet, began playing with drugs, thinking he could remain their master, but they weakened his will and sapped his energy and brought him to the grave—a defeated husk, with the marrow of his genius dried up within him.

Who made a greater appeal to his world than Byron? Who dazzled the courts and cities of Europe like this romantic? And who at any time has more poignantly expressed the bitter anguish of moral and spiritual death?

> I am ashes where once I was fire
> And the soul in my bosom is dead,
> What I loved I now merely admire
> And my heart is as grey as my head.

63

This vision of Ezekiel is more than just beautiful poetry. It is stark realism.

The second thing that meets us here is a *gospel of regeneration:* "Thus saith the Lord . . . I will cause breath to enter into you, and ye shall live." Many and ingenious are the efforts made today to inject life into the bloodstream of a dying Church. There are a group of ministers in the free churches who think a liturgical revival will bring about the miracle. Our religion is too dull and drab, too plain and pedestrian. Bring back color and movement and beauty, and the disgruntled masses will return. What incredible naïveté and deliberate running away from facts. The situation confronting us is too stark and terrible for the making of polite noises. Why not see the writing on the wall—that the churches which specialize in that sort of thing are by and large the deadest in the world, that it takes something more drastic than the repeating of the correct collect, with the correct accent, and a liturgical meandering around a chancel, to restore life to dead bones?

Liturgical Revival = Inadequate

There is another school of thought which tells us a social reorientation of the gospel message is our only hope. Those who have read much of this book cannot accuse me of soft-pedaling this issue. A religion that is merely private and individual is pathetically out of date, and those who sponsor it would feel more at home in a museum than in the modern arena. Christ is the Saviour of the body as well as of the soul, of the material as well as of the spiritual, of the realm of politics and economics as well as that of private psychology. All that is true, but the social gospelers have distorted and diluted the message by thinking you can build the kingdom of God out of men and women not personally committed. But, thank God, we have been given the gospel of regeneration, which time and again throughout history has blown its reviving breath upon the dry bones of dead things and caused them to "[stand] upon their feet, an exceeding great army."

Social Reorientation Inadequate

So it was when Ezekiel, languishing by the alien waters of

Rebirth

Babylon, proclaimed it, and a dying, defeated nation came to life, to produce in time a Paul, a Peter, even a Christ himself. So it was when in 1516 a distracted monk, reading Paul's Romans in the tower-room of Württemberg Castle Monastery, felt a great peace flooding his soul, and Luther proclaimed it to the whole world and a dead Europe came to life again. So it was in England in the eighteenth century when John Wesley, at the prayer meeting in London, felt a strange warmth creep round his heart, and he roused an astonished England from apathy and spiritual death. So it was in nineteenth-century Scotland, amid pomp and patronage, that Thomas Chalmers, an abstract mathematician, felt the wind of the Spirit stirring among the dry bones, and he arose and set the heather on fire for Christ.

And—mark this well—that does not belong to the past. It will happen again. The Christ who, that terrible night long ago, defeated death and dispelled darkness is alive today, still mighty to save. This is the gospel of regeneration. Let us proclaim it to all the earth.

The third thing that meets us here is a *gospel of resoluteness:* "Prophesy upon these bones, . . . hear the word of the Lord." "But, Lord," remonstrated Ezekiel, "the thing is impossible." "Yes, by human calculations, impossible," answered the divine Voice. "But this is a command; go and do it." This note of resoluteness has all but disappeared from contemporary religion and may be the explanation of its decay and decline. But it has by no means lost its appeal. Wherever it is sounded forth, it can still rally recruits, storm the human heart, and call forth the most passionate discipleship.

Why do you think Captain Carlsen captivated the imagination of the world? Because he possessed what men deep in their hearts admire and even worship—courageous resoluteness. We Christians have lost it. We are too conventional and cautious and calculating. We captivate no one's imagination. We rouse no soul-stirring interest, and we are simply ignored. Anatole

France once, in a mood of bitter self-criticism, said, "I have spent my life twisting dynamite into curl papers." When you think of the rebel, revolutionary faith of Jesus and the timid, shadowy, colorless thing we have made of it, it is too terribly true that we are just playing, "twisting dynamite into curl papers."

Today in certain circles it is fashionable to debunk John Knox. Certain well-bred people can't stand the great reformer and brand him as a lecherous old man, crude, cantankerous and unromantic. But these finicky and fastidious folk forget one all-important fact; the ship of reformed religion in Scotland was listing heavily with some shifty cargo aboard, and near to foundering. Who navigated "the rough seas and the fierce storms of persecution, kept it afloat and brought it safely to port"?— the man we call John Knox. Read his own brilliant work *The History of the Reformation,* edited by Professor Dickinson of Edinburgh University, and in spite of many human failings, one clear fact emerges. Here was a man of resolute courage who, like Ezekiel, was called forth to blow the breath of life upon the dry bones of a dead and corrupt church.

Do you know what hurt me most during the war? Not the catastrophic collapse of public and private morals; not the noisy crumbling of the very pillars of civilization; not the unspeakable suffering and the senseless, bloody carnage of it all; not that, but the spectacle of courageous men who died in their thousands and had no use for religion. They defeated the Nazis, laughed in the face of death, died with a fierce and gay abandon, and felt nothing but a searing contempt for the Christian church. And the reason? Because we who profess Christ are so timid and irresolute and colorless. We have taken the dynamite of the Cross and twisted it into curl papers.

Lawrence of Arabia in a brief sentence lets us into the secret of his astonishing power as a leader. "If the Arabs knew you could more than equal them in your willingness to suffer, and if need be to die, they would follow you to the ends of the earth."

Sacrifice

66

Didn't Someone greater than Lawrence once say: "If any man will come after me, let him deny himself, and take up his cross, and follow me"? And the miracle took place: men down the ages threw caution to the winds and marched to their death. They are still doing it. When will the masses take us seriously? Only when they know beyond doubt or questioning that Christ is more precious to us than our social prejudices, our hidebound traditions, and our inflated self-importance. Then, and only then, can we prophesy and the dead bones of a disillusioned age will come to life and stand upon their feet "an exceeding great army."

CHAPTER TEN

The Conquest of Cowardice

"Then came Jesus, the doors be-
ing shut."   —JOHN 20:26

IN THE UPPER ROOM IN JERUSALEM, A FEW NIGHTS AFTER
the Crucifixion, eleven terrorized men sit huddled together.
The atmosphere is full of strain and tension. Furtively they
peer out of the windows, glimpse the enemy agents standing
at the street corners, and frantically they put up the shutters.
The door is bolted, barricaded, and they speak in whispers.
Unmanned by the events of the last few days, they sit full of
perplexity.

"Art thou he that should come, or do we look for another?"
John, languishing in prison, had asked. "Yes," said Jesus, "I am
the Messiah!"

And these men believed it, and went on believing it till the
calamity of Calvary smashed their logic to smithereens. If there
was a Rational Principle, how could these things be? If there
was any purpose in the universe, why this mad and meaningless
tragedy? If there was a person called God the Father, why did
he allow Jesus to die? These men were perplexed.

Look at their faces again, and across them you see written
the word "Failure." They had set off buoyantly confident that
the kingdom of God had come to stay, but it had all proved the
insubstantial fabric of a vision, a mocking mirage in the desert
of their existence. These men had left home and friends and
families, had cut the moorings that bound them to the past,
had gaily launched out into the deep, and now the bottom had

dropped out of their universe. They sat there crushed by a desolating sense of failure.

Look at their faces again, and across them you see written the word "Fear." Here the New Testament is brutally frank. They had locked the doors, shuttered the windows, barricaded themselves in for fear of the Jews. Their names were on the black list. They were branded as traitors of church and state. For all they knew, placards stood at the street corners with the flaming words, "WANTED BY THE POLICE: FRIENDS OF JESUS." They knew only too well the fate of those who fell foul of the state. All too vividly they remembered the spurious trial, the prolonged torture, the quivering body nailed to the cross, the cry of dereliction, "My God, my God, why hast thou forsaken me?" These men were afraid.

The emotions written across the faces of these men in the Upper Room that night are the emotions written large across the face of history today. For one thing our world is sorely perplexed. The signposts that once guided humanity on its way are down. The landmarks of centuries have been obliterated. The philosophies, faith, and panaceas in which men trusted are now in the melting pot. No one dare any longer be dogmatic. H. G. Wells sums up the contemporary bewilderment in one pithy, pregnant phrase, "Mind is at the end of its tether."

For another thing, we have a corroding consciousness of failure. Western civilization, it seems, has been weighed in the balance and found wanting. The thing we call democracy, which has weathered the storms of centuries, now appears to be crumbling up. In two devastating global wars we have sacrificed millions of young lives, apparently to no purpose. As individuals, we must definitely have failed. There is the yawning chasm between dream and reality, the leering gulf between ideal and attainment. There is the fact of tarnished conscience, hopes broken beyond repair, and disappointments more bitter than words can tell. And again we live under the shadow of fear. Ever since that fateful day when Hiroshima went up in one searing, blinding flash, we have been fear-ridden. The number one tyrant of the world today is not the atom, but fear. It is fear that forces us to spy on one another. It is fear that breeds suspicion and hatred and hostility among the nations. It is fear which one day will plunge us into war and mass suicide. Yes, the scene in the Upper Room that night is terribly relevant for ourselves and for our world. So what I want you to do is to follow the text in its inner significance. It isn't just ordinary walls that shut men in. It isn't just fear of outside forces that make us lose our nerve. It's secret doubts that wall us up. It's inner fears that unman us.

All of us find ourselves shut in behind the impenetrable iron curtains of life.

THE CLOSED DOORS OF FAITH

Most of us are tired of hearing that our age is secular, that our world is grossly materialistic, that our generation has lost all capacity for faith and the will to believe. That there is a great element of truth in this charge no one who is aware of the appalling spiritual apathy of our day can possibly deny. But it is only half the truth. The other half is that men really want to believe, but they come up against a closed door against which they batter their heads in vain. You would be surprised at the number of men and women, ostensibly worldly and cynical, who in secret have knocked and hammered and battered at this shut door till at last in sheer despair they have given up. No one would call Somerset Maugham a religious man. Outwardly he is clever, worldly, sophisticated. A streak of undisguised cynicism runs through all his writings. Yet in his autobiography *The Summing Up,* one discovers with surprise that this man has been haunted by religion all his days. He had studied philosophers like Plato and Plotinus; theologians like Paul, Augustine, and Aquinas; mystics like St. Teresa and St. John of the Cross. He had read more widely in this field than the average ordained clergyman. For long years he had hammered importunately at this closed door. Desperately he had tried to tear down the iron curtain that stood between him and God, without success.

Hunger for Faith

Bleakly he sums up, "There is nothing." Maugham there is speaking for countless numbers. We are apt to treat Omar Khayyám very lightly, to dismiss him summarily as an unredeemed sensualist, a born epicurean addicted to wine, women, and song. But there is one verse in the *Rubáiyát* which shows that he was a man who had battered at this iron door for many years:

> Myself when young did eagerly frequent
> Doctor and Saint, and heard great argument
>   About it and about: but evermore
> Came out by the same door where in I went.

There may be some people who will read this who, lacking Maugham's ability and Khayyám's genius, yet in their own way have hammered at this closed door of faith all their days and are now thinking of giving up in despair. In anguish, they keep crying out, "Why is the door so stubborn and unyielding? Why is the iron curtain so baffling and bewildering? Why is the veil so solid and impenetrable?" The reason may be that we are trying the impossible task of fathoming the divine mystery by our own puny minds, of tearing down the separating veil by our own cleverness, and in our preoccupation we have forgotten the most important fact of all—that Christ is ALIVE and that if we really want him no closed door, no iron curtain, no separating veil will ever impede his coming.

THE CLOSED DOORS OF LATENT POSSIBILITIES

In a novel of Koestler's there is a vivid scene centering around a character called Leontiev, a writer who renounced Communism because he wanted to write as he pleased. He sits before a desk, slides his fingers tenderly along its polished surface. In the center lies a sheaf of soft, creamy paper, and with poised pen he sits, dreaming of the various chapters he is about to write. There was to be a chapter of scorching indignation against the crying injustices of the earth, a chapter of flaming passion to restore truth to a lost and despairing world, a chapter of broken-hearted lament as if written by the waters of Babylon, but when he tries to write nothing will come. All he can put on paper are the words on the top of the page, "I was a hero of Culture." Instead of a liberating stream, all he could squeeze out of himself were these mean, prickly, acid drips of sweat.

One of the most poignant of all tragedies is the spectacle of men, with great and glorious possibilities bottled up within them, never once discovering the secret of their release. In certain cases, something dramatic happens. The sluice gates behind which human genius is dammed up are suddenly opened and the liberating streams come rushing forth. Abraham Lincoln was a monotonous failure. He tried his hand at business, teaching, and law—in vain. Then came the Civil War, and for the first time in his life he was able to express himself. There was Churchill languishing in the political wilderness, his great genius throttled. Then came the war, and he rose to the height of his tremendous powers. You and I are not Lincolns or Churchills. We shall never be called to shoulder such mighty responsibilities, but like Leontiev in the novel, we feel our genius is bottled up within us and we have lost the secret of self-expression. We have stood helpless before the shut door of latent possibilities. We have hammered and battered and knocked importunately, we have put our shoulder against it and heaved with all our might, but nothing has ever happened.

Think again of the huddled band in the Upper Room that fateful night. In one corner you see Matthew looking scared and frightened and impotent. In another you see John, dazed and stunned and helpless, and over against the wall you see Peter, slumped in unbelievable dejection. Then came Jesus, the doors being shut, and there and then a miracle happened. Matthew's genius gave us the Gospel with which the New Testament starts; John's the fourth Gospel, and the Epistles of Love. Peter became the leader of the early Church, the rock against which the gates of hell could not prevail. And so it has always happened from age to age. Paul was a fanatical Jewish persecutor; then came

72

Jesus and he became the greatest genius in Christian history. There was Francis Thompson, whose poetic genius was choked by drink and drugs. Then came Jesus and he wrote:

> Not where the wheeling systems darken,
> And our benumbed conceiving soars!—
> The drift of pinions, would we hearken,
> Beats at our own clay-shuttered doors.

THE CLOSED DOOR OF HOPE

Take one last look at the faces of these frightened disciples, and deeper than their sense of perplexity, deeper than their sense of failure, deeper than the emotion of fear, is the look of abject, irretrievable hopelessness in their eyes. In Galilee things had been sometimes difficult, but Christ had always been there. In Jerusalem the last week was a veritable nightmare, but Christ was there. But now Christ was gone and with him faith was gone and love was gone and hope itself was gone. and the future yawned before them, grim and meaningless and empty. Then came Jesus and everything was changed. Peter cannot hide his excitement and exhilaration when he blurts out, "Blessed be the God and Father of our Lord Jesus Christ, which according to his abundant mercy hath begotten us again unto a lively hope by the resurrection of Jesus Christ from the dead."

One of the outstanding features of our age is a bleak and blank pessimism which has gripped the world by the throat. The froth of Victorian optimism has been blown away by the hurricanes of history, and a deep undercurrent of despair can be detected in the writings of the poets, the novelists, and the philosophers—in Auden, Hemingway, and Huxley; Wells, Russell, and Koestler. It began to appear first in the writings of Matthew Arnold, who saw clearly enough that when faith dies, hope dies with it. Behind the lines of his poem "Obermann Once More," you can sense this note of wistfulness:

73

When faith dies,
hope dies!

Oh, had I lived in that great day,
How had its glory new
Fill'd earth and heaven, and caught away
My ravish'd spirit too!

. . . .

Now he is dead. Far hence he lies
In the lorn Syrian town,
And on his grave, with shining eyes,
The Syrian stars look down.

Matthew Arnold is right. Without a risen Christ there can be no hope either for this life or for that which is to come. It is because we Christians believe that Christ has conquered death and is alive for evermore that we can never give way to the despairing pessimism of our times. Indeed, that prevailing pessimism becomes the herald of Christ's coming. When man can solve all problems in the world and out of it by his own cleverness, when things are not too desperate and he is able to stop on this side of his own ability, he can get on nicely without God. But when man is at the end of his tether, when his best-laid schemes have gone wrong, when every door is shut, then he is ready for the coming of Jesus.

In the last resort it is not we who open up the iron curtains behind which we are shut in. It is Christ himself who comes through them. He comes in the time of our greatest need, in the hour of our deepest perplexity.

CHAPTER ELEVEN

What It Means to Be Healthy

"They that be whole need not a physician, but they that are sick."
—MATT. 9:12

ONE OF THE MOST CHALLENGING FACTS CONFRONTING US today is that in the most advanced forms of society religion appears to be on the way out. Europe, the birthplace and the center of Western civilization, demonstrates this all too clearly. The decline of faith is usually ascribed in our own country to the shattering effects of two wars; but Sweden and Switzerland have not been involved directly in military conflict now for centuries, and the symptoms are as plain there as here. The truth seems to be that that decay had set in before the turn of this century. What two devastating wars did was to accelerate the progress of the rot that had already started.

One symptom of the general decline is the serious falling off in church attendance. In that most interesting book *English Life and Leisure,* Rowntree and Lavers supply us with frightening statistics. They select York as the center of their investigation. An archiepiscopal city, steeped in tradition and religious atmosphere, it is a place where you would expect people to be more church-minded than in industrial cities like Glasgow, Manchester, or Leeds. What did they find? In 1900, thirty-seven out of every hundred attended church; in 1948, only thirteen—roughly a third.

Another sure symptom of diminished interest is the devaluation of religious words which in the past circulated freely and were adequately understood. The word "salvation" is a case in point: Paul, writing in the first century, uses it frequently, though writing to converts only recently reclaimed from paganism. Neither the educated classes nor the working man in the brass foundries of Corinth had the slightest difficulty in understanding what he meant. The same was true when the fire of the Reformation blazed across Europe. The princes and the peasants, the cultured and the common people, knew what salvation was. It was even true when Henry Drummond spoke to crowded student meetings in Edinburgh at the beginning of this century. The future doctors, scientists, and teachers of our country were not bamboozled by this word. They might not all accept "salvation," but they knew what Drummond meant by it.

But now the word has become devalued. If you mention it in educated circles, you will find people looking at you with incredulity. As for the ordinary man, he will gape openmouthed and have no idea what you are talking about. The term has, by and large, become unintelligible.

Perhaps part of the task of modern evangelism is to get behind religious terminology to the realities which the words represent. If we read the New Testament carefully, we shall find that Jesus was not tied down by contemporary theological jargon. Quite often he broke loose from it and used an idiom that was fresh, and original, and startling. There are times when he equates salvation with health, meaning not only health of body but that of mind and spirit as well. Here, surely, is something which is well within our understanding, something which is familiar to us all, and which we can quite readily grasp and understand.

When the Pharisees accused Jesus of consorting with publicans and sinners, he replied with fine irony, "They that are whole need not a physician, but they that are sick." According to Jesus, sin was a deadly disease—a malignant growth which, like cancer, extended its roots and ate into the vitals of personality. If this is so—and we have Christ's own authority behind us in claiming that it is—salvation is never out of date, never obsolete or irrelevant. It is, in fact, imperative.

MANKIND ON THE SICK LIST

Physically we may be bigger and healthier than our fathers, but, generally, we are far from well. Never in all its long history did our world have within its grasp such golden opportunities. The airplane, the radio, and television have conquered space; natural boundaries have become artificial and arbitrary; science itself has showered upon us boons and blessings inestimable and could open up an era of unparalleled comfort and happiness. Yet we have behaved like maniacs let loose. The Nazis are supposed to have exterminated eighteen million people, using every conceivable refinement of brutality and torture. In a few searing seconds 200,000 people in Hiroshima were instantly incinerated, mutilated, or doomed to die. No man in his senses can call our world healthy. It is sick with a sickness that appears to be incurable.

Our writers have diagnosed this sickness and in plain, unvarnished language tell us there is no cure this side of time— "we are doomed to die."

Jung, the famous psychologist, who has treated literally thousands of the mentally sick from all over the world, says in his book *Modern Man in Search of a Soul* that those who came to him for help are in fact victims of a deep-seated spiritual malaise for which the real cure is not psychological but religious.

Lewis Mumford, the brilliant American sociologist, sets down a very terrible thing in his latest book, *The Conduct of Life:*

My own experience as a teacher . . . makes me believe that perhaps as much as a third of our student population of college grade may, for all practical purposes, be considered moral imbeciles, or at least moral illiterates.

The observations of these intellectuals have been endorsed by my own personal experiences. The most all-round brilliant man I met during the war—a first honors graduate of Oxford, a superb fighter-pilot, a sportsman of international standard—I would unhesitatingly describe as a moral moron.

And, if we are honest with ourselves, we must confess that we are all smitten in varying degrees with the same disease. There is an impassable gulf between our ideals and our attainments—a deep disharmony within which plays havoc with the music of life, a principle of contradiction which makes us cry out: "The good that I would I do not: but the evil which I would not, that I do." We understand what Christ meant when he talked of sin as a deadly species of sickness. We are all sick and we are desperately clamoring for a cure.

THE POPULAR REMEDIES

Modern man, though not crushed by a consciousness of sin, is prepared to admit that something is wrong. Something, he is beginning to see, has impeded his progress and sabotaged his designs—something has thrown a monkey wrench into the works.

He has not hitherto been at a loss for a reason. Many slick solutions are invariably at hand—panaceas in which he has placed implicit confidence. There is science, ready to harness the hidden powers of nature; education, ready to dispel our ignorance; social reform, ready to banish our injustices; psychology, ever ready to dissolve our complexes.

Well, listen not to a theologian with an axe to grind but to a scientist. Sir Sydney Smith, professor of forensic medicine in Edinburgh University, in a recent lecture claimed that if the conditions of life were the determining factors, crime would now be on the decrease. In the last fifty years we have seen a fall in the death rate, a rise in the expectation of life, and a much reduced infant mortality. We have gained control over many diseases and learned much about nutrition. Among social changes are ensured high wages for the working man and less unemployment; while primary, secondary, and university education are within the reach of all able minds. We have offered new opportunities to young people in sport and pastime. The movies and radio have opened up new doors of learning and enjoyment. The treatment of offenders has been progressively more enlightened. The emphasis now is reform rather than retribution. Yet, in spite of all this, things, far from improving, have grown progressively worse. Serious crimes like murder, rape, assault, and larceny have multiplied, and juvenile delinquency is on the increase. Sir Sydney ends up pointedly by saying, "Crime has not been favorably affected by attempts at reform or rehabilitation." In other words the easy solution, the patent medicines, the proffered panaceas have all failed, and if man is to be saved he must look in another direction.

CHRIST THE GREAT PHYSICIAN

They that are whole need not a physician, but they that are sick. And what a Physician he was! No modern doctor has shown greater interest in the bodily health of men than Jesus. From dawn to dusk he toiled at his healing work—the blind, the deaf, the paralyzed, the palsied. But that was not all: he took the cynical, the disillusioned, the defeated and despairing, the mentally distraught, the nervous wrecks of life and changed them into new creatures—purposeful, forward looking, bursting with new zest and energy.

He took Mary Magdalene, the woman of the streets, and transfigured her into an ideal of beautiful womanhood. He took Zacchaeus, an unscrupulous twister, and made him into an honest man. He took Peter, a groveling coward, and made him into a saint and martyr. He took Paul, a hate-filled fanatic, and made him into the apostle of love. There was no disease this Physician could not cure, and today his skill is not spent. His touch has still its ancient power.

His cures are not confined to the New Testament. They are spread over the centuries. He took Augustine in the fifth century and cured him of pride, intolerance, and sexual passion. He took Luther in the sixteenth and cured him of melancholy and inner conflict. He took Francis Thompson in the twentieth, and cured him of drink and dope.

Kagawa of Japan was ridden with disease. He had tuberculosis in both lungs. He had hemorrhage after hemorrhage. He also had trachea, heart, and liver trouble, and he once had the unique experience of seeing a doctor write out his death certificate. But this frail little man is now a veritable giant. His record has been terrific—minister of agriculture and fisheries, member of the cabinet many times, evangelist, theologian, builder of a thousand churches, author of over sixty books, social reformer, politician, revolutionary. How did he do it? Well, listen to his own testimony:

I owe everything to Christ. I owe Him my physical life, for without Him I would have died as a young man of T.B. All that I am and that I have accomplished I owe to Him, for He took me for His—to care for me, a young man—naturally timid—full of fears, and He made me a disciple prepared to suffer, to bleed and to die that He might be exalted among Men.

CHAPTER TWELVE

The Secret of Success

*"When they saw the boldness of
Peter and John . . ."*—Acts 4:13

BEVERLEY NICHOLS, IN HIS BOOK "A PILGRIM'S PROGRESS,"
gives us a pen portrait of a typical summer Sunday morning
in London. As he approached the city from the south, he found
the main road clogged with a roaring traffic of cars, buses, and
cyclists speeding toward the coast. He parked his car near a
large church and sat behind the wheel, watching. At a quarter
to eleven, two old ladies walked up the steps with some difficulty. For three minutes nothing happened; then two more
elderly ladies appeared, one of them leaning on a crutch. Just
before eleven the rest of the congregation arrived; it consisted
of nineteen elderly ladies, seven old gentlemen, and an angry-
looking nurse dragging a reluctant child by the wrist.

He is deliberately exaggerating to dramatize his theme, dis-
torting the facts to heighten the pathos of it all. But it brings
home to us the shattering, soul-destroying indifference so grimly
characteristic of the religious life of Britain today. While our
world is stumbling from one crisis to another, and our own
country is passing through times that make the utmost demands
on all our latent resources, men and women, in increasing
numbers, turn their backs on worship and seek out some
favorite haunt—to lie on the sands or loll in the sun.

No wonder so many preachers get tired and discouraged.
The gospel is regularly proclaimed from pulpit, press, and radio.
There is no scarcity of new experiments, novel approaches, and
revolutionary ideas; but in spite of it all, the churches remain

depressingly empty on Sunday while packed pleasure-coaches speed from John o'Groat's to Land's End. If we can judge by appearances, the masses are appallingly apathetic. They couldn't care less!

Scotland is one of the smallest countries in Europe. Its national church still commands respect and enjoys a considerable measure of prestige. She sits every year in general assembly, and her deliverances on such problems as Communism and racialism are often more intelligent and revelant than those of the House of Commons in London, or Congress in Washington. Why then does she not carry more weight? Why do men go about their business heedless of her call?

Therein lies the challenge of the early Church. Not so well organized or well informed, nor nearly so strong numerically as we are, she made an impact on her age out of all proportion to her size. The opposition drove her to the catacombs, determined to crush and trample her out of existence, but they could not silence her. These first Christians took Jerusalem by storm. So strong was the interest roused that the leaders could no longer turn a blind eye to it. Hasty meetings were convened and with astonishment they marked the boldness of these flaming revolutionaries. Why were ordinary men like Peter and John able to make a callous, cynical world sit up and take notice? I suggest the reasons were:

THE BOLDNESS OF THEIR CREED

After the lapse of long centuries it is difficult for us to understand how deeply those disciples startled their contemporaries. This is in effect what they proclaimed: "The man Jesus whom you tried and destroyed was none other than the long-promised Messiah. You crucified him and buried him deep in a tomb, but he has defeated death and risen and triumphant now reigns as King of kings and Lord of lords. This is no hearsay: we who speak have seen and handled him, and the Holy Spirit he promised to send came on the day of Pentecost like a rushing mighty wind and cloven tongues of fire and took possession of us." To those who listened, this sounded preposterous—the most absurd of all possible illusions. The brazen boldness of the claim took their breath away.

Ever since the Renaissance men have been trying to water down the Christian creed. Give us a religion purged of everything that defies logic, a religion stripped of the supernatural and emptied of miracle, a religion that is smooth and palatable and rationally acceptable: this has been the popular cry. And Liberal Protestantism went out of its way to oblige. The contempt in which the church is held by so many intellectuals today is largely the result of such shortsighted concessions.

Not that we who believe despise the intellect. Reason is the gift of God, and a religion that does not appeal to it has nothing to say to an age steeped in a skeptical scientific atmosphere. Theology is necessary, not only to preserve the faith from the benighted superstitions of the more crazy sects, but also to preserve modern culture from intellectual suicide. But the emphasis must never be placed on conformity to contemporary thought categories. It must be placed on the soul-splitting facts of the gospel in the incarnation of Christ, the cross, and the power of his resurrection.

I once heard C.E.M. Joad address a large gathering in the St. Andrews Halls, Glasgow. He pleaded with theologians and preachers not to come to too easy terms with modern culture. "I know my generation," he said, "and if you are to convince them you must confront them with a bold, authoritative, and demanding creed." "Mind is at the end of its tether," cried the disillusioned rationalist, H. G. Wells, at last becoming aware of the bankruptcy of all rival substitutes. To make such men sit up and take notice we must declare a positive gospel which demands not merely an intellectual assent but a total response on all levels of personality.

THE BOLDNESS OF THEIR CONDUCT

The commotion in Jerusalem that day was not caused by mere words. These apostles were prepared to back their claims with their very lives. And the question on everybody's lips was, "What has happened?" It was commonly known that these men who now preached with such reckless abandon had only yesterday deserted Christ in his hour of need. Had not Peter denied him with loud, blasphemous oaths in the courtyard of the procurator's palace? Yet now, within a stone's-throw of this scene of shame, he was proclaiming him with intense and passionate conviction. No wonder the scoffers sat up and took notice.

And here history has repeated itself, not once but a thousand times. Whenever men appear who eschew calculating prudence and throw caution to the winds and scorn all consequences, the world sits up and takes notice. Einstein had regarded the Christian church with polite contempt, until he saw the stand she made in Hitler's Germany. When the Nazis formulated and proclaimed their satanic creed, who was it who stood in the way? Not the universities, nor the newspapers, nor big industrial combines; they soon crumbled and conformed. Only the church of Christ fought back and went to the wall rather than surrender, and, looking on, the cynics and spectators sat up and took notice.

Today as never before we need that brand of courage which remains true to Christ at all costs. Here in Britain we may not understand the South African situation in all its maddening complexity, and we should speak therefore with studied restraint. But there can be no doubt that some of the policies pursued are far from Christian. What are ministers of the gospel and professing believers to do? Are they to remain silent and conform to the *status quo?* Or are they to preach boldly in word and action the social and political implications of the gospel Christ came to declare unto all men?

Again, on this side of the ocean, we do not always grasp the intricacies of American politics, and we should rightly hesitate before we foolishly criticize. But I do not see how the American church, at the moment excelling in great vitality, can evade this question—What action am I to take when demagogues like

McCarthy deify the American way of life and, in the process, tamper with conscience and interfere with liberty?

And, anyhow, we British Christians can't afford to throw stones at anyone. We have lost the bulk of the working classes because, in the past, the church remained silent while they were exploited by a cruel and ruthless capitalism. We still cringe, silent and stricken, before national iniquities that are truly monstrous, and we are afraid to act because the Cross has become a vague, theological concept, not a sword in our bones that drives us to costly action.

THE BOLDNESS OF THEIR COMMISSION

"Go ye into all the world, and preach the gospel," commanded Christ, and from the start it looked pretty suicidal. The odds against the disciples were overwhelming—the implacable bitterness of an intolerant church, the crushing weight of a predominant paganism, and the mailed might of Caesar's conquering legions. If you and I had been there, we would have written off the whole thing as a sorry fiasco—"but by A.D. 323 the impossible had happened." Constantine made Christianity the official religion of the far-flung Roman Empire.

And whenever men have marched out with a bold and impossible commission, a similar miracle has always happened. The Reformation looked dim and remote, a futile and forlorn hope, when pessimistic friends counseled Luther not to go to the Diet of Worms. "The Pope and cardinals have signed your death warrant," they said. "I don't care," answered the reformer. "The princes are in league against you, and they are plotting your destruction," they pressed. "It matters not, I am going," said the stubborn monk. "The Devil is lying in wait for you," they cried, playing their trump card. "Let there be as many devils as there are red tiles upon the roof tops and they won't prevent me," answered the great committed soul. And more than half of Europe marveled at his boldness and sat up and took notice.

And the secret of this conquering courage? It is found in

these simple words: "They were with Jesus." That is the Achilles heel of modern evangelism. Our diagnosis is invariably clever and quite frequently correct. Our equipment is superb, and the pattern we seek to impose on society Christian and realistic, "but at the most crucial point of all we fail lamentably—because we do not know Jesus."

This Christ-centered consciousness is basic. We cannot build the new Christian order out of men and women who have not known Jesus as a personal Saviour. We cannot generate Christian courage self-consciously or synthetically. It must flow spontaneously out of a personal allegiance to Christ, for "he that findeth his life shall lose it: and he that loseth his life for [Christ's] sake shall find it." That, paradoxically enough, is the only way we can make the world sit up and take notice.

CHAPTER THIRTEEN

The Inner Nerve of Anxiety

*"Let not your heart be troubled:
ye believe in God, believe also in
me. In my Father's house are many
mansions."* —JOHN 14:1-2

WE ARE ALL IN VARIOUS DEGREES FAMILIAR WITH ANXIETY.
Looking back, I can vividly recollect some anxious moments
in my own life—the first sermon I preached, the first bereaved
family I visited, trembling and tongue-tied, the first parachute-
jump I made from a balloon. These are the sharp, terrifying
experiences I shall never forget. But anxiety I also know in its
less dramatic but more chronic forms. Once a year at least I
have a ghastly nightmare when I see my congregation, disillusioned with me as their minister, disintegrating before my
eyes, and I wake up trembling. Though no psychoanalyst, I
think I know the explanation. The anxiety, which on the
conscious level I don't succumb to, lurks hidden in one of the
deep caverns of the subconscious, and when I am overtired,
off guard in my sleep, it launches the savage attack. And so,
my friends, I can preach on this theme without going one iota
beyond my own experience.

No doubt your anxiety will take different forms. You may
be anxious about your work, your health and family, or, if you
are a student, about your career. It may be even more inti-
mate—the consciousness of your own inadequacy to meet the
pressing demands of life. Here we are all vulnerable. There are
no exceptions. The debonair, devil-may-care pose is often only
a mask. It fails to hide the fact that in the secret places we are
all desperately anxious.

SOURCES OF ANXIETY

A great deal has been written on the subject, especially by modern psychologists of note—Freud, Adler, Jung, McDougall, to mention only a few—but as far as I can see, not one of them has touched the inner nerve of the problem. By far the most profound work I have read on this subject is by Paul Tillich, now professor of theology at Harvard University. In an exceedingly difficult book which compels the reader to perform mental somersaults, he traces anxiety to three main sources.

The first of these may be called *a sense of meaninglessness.* This can be illustrated from literature much more effectively than from Tillich. The most forceful expression of this sense of life's utter futility I have encountered is in Tolstoy's *Confession.* At a time when he was at the pinnacle of his power as a novelist, with a world-wide reputation, enjoying splendid health and congenial work, he suddenly lost all desire to live. Hunting, playing with the children, working on the land—things he normally loved—ceased to interest him. His energies were short-circuited and he more than once contemplated suicide. Recent literature, poetry, drama, and the novel abound with examples. A sense of "meaninglessness" goes hand in hand with anxiety, sometimes bordering on despair.

The second is *a sense of guilt,* and perhaps the best portrayal of this is to be found in Sophocles' famous play *Oedipus Rex,* about the son of the king and queen of Thebes who was removed from his parents at birth and who later on unwittingly slays his father and marries his mother. When at length he discovers the awesome truth, his sense of guilt is so overwhelming that he puts out his own eyes and condemns himself to eternal exile. It is an old story, but it reminds us that we cannot ignore guilt. It haunts us and hounds us, tracking us down in the end and playing havoc with our peace of mind.

The third—*the fear of death*—is, according to Tillich, the basic source of all anxiety; the fear that the being which is I will cease to be. If you counter that this is not true, that the

88

recipients of the Victoria Cross prove their utter contempt of death, I can only answer that, in point of fact, it proves the contrary. The reward of the Victoria Cross is a recognition on the part of society that the fear of death is the basic anxiety.

SO-CALLED PALLIATIVES FOR ANXIETY

These then are the sources of anxiety, and now comes the question: What is the remedy? How can we alleviate and allay this inner tension that so often wrecks our happiness and destroys our peace of mind?

There is the answer of the poet, or at least of a certain type of poet, who offers nature as the cure:

> The little cares that fretted me,
> I lost them yesterday,
> Among the fields above the sea,
> Among the winds that play,
>
> Among the lowing of the herds,
> The rustling of the trees,
> Among the singing of the birds,
> The humming of the bees.

This may dislodge a trivial worry, but it leaves the basic problem untouched.

There is the answer of the sentimentalist—the type of person who never faces unpleasant facts. Sometimes they are found among the ranks of the psychologists, those who sneer at puritanism, who label guilt as the stock-in-trade of a morbid theology, who urge us to remove the moral hatches and let the damped-down passions come up to the bridge of personality. Some of them are ministers whose sole concern it is to make people happy (when, if they were true to Christ, they should make some people desperately unhappy). Their theme is relaxation. A theological professor bitingly parodies one such popular preacher in America whose gospel is—"Go ye out into all the world and relax!" They offer people a technique, a theological pill, a soothing anodyne which they swallow in order to ease the inner strain. They mutilate Christ's gospel by emphasizing the comfort of religion at the expense of its challenge, and the only way they can enable people to relax is by persuading them to blind their eyes to the grimmer side of reality and asking them to live in a fool's paradise.

The truth is that it is never possible this side of eternity to overcome anxiety completely. It is a mark of our creatureliness, a sign and symbol of our finitude. I think I know what mood Walt Whitman, the poet, was in when he said: "I think I could turn and live with animals; they are so placid and self-contained. They do not lie awake at night and fret over their sins." However much we may aspire to, we can never quite achieve such bovine placidity. Augustine was much nearer the mark when he cried: "O God! thou hast created us in thine own image, and our hearts will ever be restless until they find rest in thee." There is a very real sense in which Christianity adds to, rather than subtracts from, our anxiety. It enlarges the area of our concern, sharpens our sympathies, makes us more sensitive to the sufferings of our fellows. How can anyone relax in a world where races are cruelly segregated, and displaced persons without homes, jobs, or civil rights are herded like cattle in camps behind barbed wire? How can anyone relax who takes seriously the Christ who said: "I came not to send peace, but a sword"?

CHRIST AND ANXIETY

But that is preposterous, you say. What a depressing, dismal message! Do you mean to say there is no remedy for anxiety, no final solution, no secret of resolving this inner stress? Yes and no is the honest answer. No—because we are finite creatures, made in God's image. Yes—because Christ has given us the key enabling us to see anxiety in its proper perspective.

The scene in the Upper Room on the eve of Calvary is by far the tensest in history. In comparison, the issues hanging in the balance on the eve of the Normandy invasion when the order to invade was about to be given were relatively unimportant. The shape of things to come, the existence of a Christian civilization, the salvation of the world, depended on what was happening in Jerusalem that night. Outside, the night was pregnant with evil machination and sinister possibilities. Death hung in the air; crucifixion was as certain as tomorrow's rising sun. To suggest that Jesus was not anxious is sheer sentimental absurdity. The figure who was soon to be seen praying in Gethsemane, with beads of bloody sweat on his brow, "Father, . . . remove this cup from me: nevertheless not my will, but thine, be done," knew the meaning of anxiety as no one else ever did. How then explain the serene calm with which he addressed his terrified disciples in words that have become immortal—"Let not your heart be troubled: ye believe in God, believe also in me. In my Father's house are many mansions"?

There you see in action what we call the paradoxical miracle of grace. Jesus, within hours of crucifixion, pushed his own anxiety into the background, because he was absorbed in two things—the comfort of his disciples, and communion with God the Father—and there is no other answer. Only in the measure in which we lose ourselves in serving our fellows and our God do we achieve serenity. Any other peace is a fantasy and a delusion.

ANSWERS TO ANXIETY

Earlier on in this sermon we mentioned the three sources of anxiety—meaninglessness, guilt, and the fear of death. Curiously enough the answer Jesus offered the disciples that anxious night in the Upper Room was threefold. We might call them the three nerve centers of inner tranquillity.

The first is *You believe in God.* That and that alone can put to flight the sense of meaninglessness which undermines the will, numbs the feelings, paralyzes the energies, and robs a man of inner peace. Belief in God invests life with a sure purpose and provides us with a final goal toward which we can strive with all our might and main.

This happened one day to Thomas Carlyle while here in Edinburgh. He was passing through a very anxious time, sleeping badly, unable to concentrate, because he was not sure that his existence had any point or meaning. While walking down Leith Walk he had an experience of a kind that can only be called mystical. He asked himself the reason for the obscure and pusillanimous apprehensions which he continually felt and, in his own words, "In that instant the Everlasting Yea rose up to choke back the Everlasting Nay." From that moment, many of his fears dissolved and he was able to commit himself to purposeful action.

The second is *Believe also in me.* Carlyle's tragedy lay in the fact that he never got beyond a belief in an Everlasting Yea— a positive purpose, a remote and abstract God. The genuinely converted Christian believes not just in God but in a God who became incarnate in Jesus. Belief in purpose is necessary if you are to allay anxiety, but it is not enough. You may have heard of the mother who was trying to train her small boy to sleep in a room all by himself. He protested and cried bitterly. To console him she bought him a brand-new Teddy bear and said to him, "Now you have a companion, you won't be lonely." He sobbed out his pathetic answer: "Yes, Mummy, but he has no skin on his face!" A profound parable. This is precisely what Jesus did. He took the principle of rationality, positive purpose, Carlyle's Everlasting Yea, the abstract God of philosophers, and put skin on its face. "Believe also in me." This is the answer to the second source of anxiety—a sense of guilt. If God is like Christ, our sins are forgiven. The broken relationship is restored, and we can experience the liberating grace of a new beginning.

> He breaks the power of canceled sin,
>   He sets the prisoner free;
> His blood can make the foulest clean;
>   His blood availed for me.

And the third is *In my Father's house are many mansions.*
The answer to the basic source of anxiety—fear of death. It is
not the crisis of death itself that we fear, but the thought that
the ego itself, the being that we nourished and developed, may
be snuffed out, crushed, obliterated. Thomas Huxley, the famous agnostic, was so depressed by this thought that once he
cried, "I would like to live on, even in hell, if only I could
know."

In the last resort, a sense of ultimate security is the only
thing that can alleviate and allay anxiety, and our sole guarantee
is the Christ who, when desperately anxious himself, spoke
words of comfort to fear-crazed men, and who came back from
the dead to prove this was not fancy but fact. If we believe this,
we can sing with Whittier:

> And so beside the Silent Sea
> I wait the muffled oar;
> No harm from Him can come to me
> On ocean or on shore.
>
> I know not where His islands lift
> Their fronded palms in air;
> I only know I cannot drift
> Beyond His love and care.

CHAPTER FOURTEEN

God and Human Suffering

*"But God commendeth his love
toward us, in that, while we were
yet sinners, Christ died for us."*
—Rom. 5:8

"AND THE KING WAS MUCH MOVED, AND WENT UP TO THE CHAM-
ber over the gate, and wept: and as he went, thus he said, 'O my
son Absalom, my son, my son Absalom! would God I had died
for thee, O Absalom, my son, my son!' "—A cry wrung from
the heart of a man hit by the sledge-hammer blow of bereavement some three thousand years ago.

Suffering is as old as that and down the long ages has haunted
men and clamored for an explanation. It haunted the ancient
Greeks. It was their preoccupation with this problem that
produced the great tragedies of Sophocles and Aeschylus. It
haunted Shakespeare. Some authorities claim that it was after
a shattering nervous breakdown, when he had plumbed the
depths of despair, that he wrote *King Lear, Othello,* and *Hamlet.* It haunted Thomas Hardy. Who can forget the heart-
rending scene in *Tess of the d'Urbervilles* of a distraught girl
dying among the eerie ruins of Stonehenge, or who can stop
his ears to the bitter cry—"Justice was done and the President
of the Immortals has finished his sport with Tess"?

The literary giants may dramatize this problem with power
and pathos, but it is by no means confined to this select coterie.
Ordinary men and women, though less articulate, feel it with
equal intensity. Aware of this cruel contradiction leering at
them from the very heart of life, they can only cry out, "My
God, why?" You remember the story of Josephine Butler, the

94

Josephine Butler

great social benefactress. One evening she was out at a party, and eagerly awaiting her return was her only daughter Evangeline. The little girl, on hearing her mother's carriage-wheels crunching the gravel, leaned over the balustrade, lost her balance, and in a second lay dead with a broken neck at her mother's feet. Josephine Butler could not speak, she could not pray, she could not cry. She stood there stunned, paralyzed by grief, muttering silently, "My God, why?"

I remember in a Tunis hospital lying in a bed next to a fellow who was mortally wounded—an English soldier who kept the proverbial stiff upper lip and never moaned aloud though there was a lamentable shortage of drugs to kill the pain. The night before he died he started talking, sensing, no doubt, that it was this last. While he spoke I watched his face, sharpened by fever and days and nights of cruel suffering. There was sweat on his brow and tiny globules glistened on the down on his upper lip. I have forgotten most of what he said, for he talked a long time, but one question he put sunk into my brain and I shall never forget it. "How can you believe in a God who lets a fellow die in hellish pain?" I have heard this ultimate problem more accurately, more neatly, more poetically, but never more forcibly expressed. Yes, indeed, the question that perplexed that dying soldier in Tunis haunts us all. How can we believe in God who subjects his children to hellish pain?

I know that I speak to men and women who have suffered, who have experienced something of what John Keats called "the giant agony of the world," and have hurled an accusing cry toward heaven—"My God, why?" I am speaking to others who so far have not tasted its bitterness. You are young and happy and full of radiant vitality. Broad sweeping avenues of hope and opportunity beckon you on. Your prospects are bright—your skies clear and unclouded. You may feel that

> God's in his heaven:
> All's right with the world.

But remember this one thing. Faith is never a matter of health and happiness. It can use these as allies, but it can operate in spite of them. The faith of Christ was never meant for blue skies and calm seas and favorable breezes. It was meant for stern days, for times of crises and convulsion, for dark moments when the floods of adversity overcome our souls and from broken hearts we call out, "My God, why?"

To this problem certain answers have been given.

The Answer of the Skeptic

Why not be realistic? Why not frankly face this brute fact of existence, argues the agnostic? Behind this vast and mysterious universe there is no creative spirit, no intelligent mind, no loving heart. It just happened as a result of a fortuitous combination of atoms; and as it kept on expanding, unpleasant things were bound to happen. You cannot win a battle without casualties. You cannot run a modern factory, however many safety devices you have, without a certain number of accidents. And you cannot live in a gargantuan universe without getting caught in the wheels at times. The best way is to accept the situation as it is and quit our wishful thinking.

It sounds brave and realistic, but in actual fact it is the bleakest and barrenest of all possible philosophies—at best a religion of "grin and bear it," a pathetic attempt to be logical in a world which is crassly and completely irrational. Think of someone quite dear to you struggling in the swirling floods of tragedy. Would you go up to him and quite calmly remark: "Don't be unduly perturbed; put your problem into its proper perspective. Realize, man, that you are only an unimportant cog in a meaningless, mechanistic universe, and, as far as the Life Force is concerned, your suffering is of no account"? Would you have the heart to say that to someone who is really hurt?

The Answer of the Spectator

Suffering is one of the facts we must take into account in life, but there is nothing we can do about it. We can only look on.

This is the sort of gospel preached by H. G. Wells in his book, *God the Invisible King*. God is good and kind and incredibly well meaning, but utterly helpless. He is like a neutral in time of war—bound by the canons of a cosmic convention; he may sympathize, but he cannot participate. He is like a spectator watching a thrilling game of football. However much he may want to enter into the game he must remain on the side lines. Curiously enough, some writers and philosophers seem to derive a measure of comfort from such a belief, but for the life of me I cannot see how. Skepticism, in a sense, I can understand, but the spectator God I can never come to terms with. What earthly use is there in a God who sits on the grandstand watching the game from a distance? As a creed, it is more blasphemous than atheism.

The Answer of the Sentimentalist

Mary Baker Eddy, the founder of Christian Science, denied the reality of pain and suffering. She suggested it was only an illusion—a mere subjective fantasy—a belief that was mentally conditioned and had no basis in actual fact. She went further than that and claimed that Christ himself had not suffered, which, if one were to take it seriously, would explode the whole scheme of Christian salvation and render the Cross absolutely superfluous. Sentimentalists like Mary Baker Eddy tell us that when we think we have toothache the pain has no objective reality. It is only a mentally conditioned experience. Instead of going to a dentist we should sit down and pray about it. I believe in prayer, but when I suffer toothache I feel I am spending my time more profitably in the dentist's chair than on my knees. To tell a man devoured by cancer, a victim of a nervous breakdown, or someone caught in the fell clutch of circumstances, that it is only a fantasy conjured up by the imagination—that is not only an insult to the intelligence, it is a cruel flaunting of all our finer sensibilities as well. I do not see how any sensitive soul can ever think of sponsoring it.

None of these answers, then, can throw any light on the unaccountable enigma of human suffering. Where are we to seek for a clue to unravel the mystery? Where look for a guiding light to scatter the shadows and show us the way? The only light, curiously enough, comes from the Cross. That is why Paul, with his paradoxical insight, concentrates on it and goes straight to the heart of things when he cries, "God commendeth his love toward us, in that, while we were yet sinners, Christ died for us."

The apostle was a profound thinker. If he had wished he could have embarked on a learned disquisition on the ultimate dilemma. He was a brilliant theologian and could very well have discussed it all on a high intellectual plane. What is so shatteringly convincing is the simplicity of the argument he employs. No philosophy, no theology, no clever dialectics, just a statement of simple fact which an untutored child can easily grasp. "God commendeth his love toward us, in that, while we were yet sinners, Christ died for us."

For you see the logic that lies at the back of this triumphant declaration—God became incarnate in Christ and Christ died for us; therefore God is no cosmic spectator, but a sharer in our agony! That places things in their proper perspective. When we suffer, what do we want? A logical explanation? A slick formula? An answer that satisfies the intellect? Certainly not! We want the comfort of assurance that when we go through some dark valley of the shadow, God is with us; his rod and staff are there.

When some sudden, unexpected calamity crashes into our lives, when tragedy hits us with a sledge-hammer blow, when we are dazed and stunned and bewildered—what is it that helps us most? Not our books or learning, or clever answers, but the knowledge that there is some dear, stanch friend ready to stand by us and share our grief. That lightens the burden, lifts the shadows, and makes life more tolerable! How much more so if we know that God himself is with us, bearing our sin, carrying our sorrows, and pressing close to us in the pain and travail of our existence!

A very difficult part of my work as a minister is to try to comfort those who have suffered. When, for the first time, I visited a bereaved family and felt painfully tongue-tied, I comforted myself with the foolish notion, "Ah, no doubt it will be easier the more experience I gain!" But now I feel as awkward as ever. The better I know them and the more I love them, the worse it gets. Now I never try to explain, for this enigma is beyond explanation. I can only reiterate Paul's supremely simple logic: "God commendeth his love toward us, in that, while we were yet sinners, Christ died for us."

As a boy, I used to wonder why our old minister at home— *Cross* a great saint he was—never preached a sermon without referring to the Cross. He might start at New York, but always he finished up at Calvary. Why didn't he give the Cross a rest, why always hark back to this tiresome theme? But now I know the answer. You cannot preach the gospel and give the Cross a rest, for it stands in all its gaunt starkness at the very heart of things. It changes everything to know that God is in Christ and that Christ suffered. The thirst he felt when he cried out for a drink was real thirst. The blood which appeared on his brow in the Garden of Gethsemane was real blood. The nails that were hammered into his body were ordinary, brutal nails. That terrible cry of dereliction—"My God, my God, why hast thou forsaken me?"—was not a piece of play-acting, but the cry of a tortured soul who carried in one terrible moment the whole of the concentrated terror and anguish and despair of mankind.

How could I ever preach a saving gospel if I could not authoritatively tell you that Christ who was God suffered and died for us? Believe that, and everything is changed. Tragedy, disappointment, defeat become not so much problems as privileges. It lifts the burden on our spirits, takes the hurt look out of our eyes, and enables us to face tomorrow with a greater measure of equanimity. This text we must learn by heart. We must repeat it till the meaning and message sink into the very marrow of our being: "God commendeth his love toward us, in that, while we were yet sinners, Christ died for us."

CHAPTER FIFTEEN

Argument with an Atheist

"And be ready always to give an answer to every man that asketh you a reason of the hope that is in you." —I Pet. 3:15

SOME TIME AGO A SERIES OF TALKS WAS GIVEN ON THE RADIO by Fred Hoyle, a young scientist. Later they were published in a little volume called *The Nature of the Universe*. This book is interesting, fresh, and stimulating, and as far as I am able to judge, intelligent. Then, in the last few pages, the author mars his work by falling victim to the disease known as the fallacy of transferred authority. He makes the mistake of transferring his authority in the realm of science to that of religion, a subject on which he is not at all competent to speak. It is a common modern malady. Brilliant men, quite humble in their own spheres of physics, philosophy, literature, and art, become little oracles and speak with shattering omniscience on the mysteries of the faith. The ordinary man, mesmerized by science and mystified by religion, takes his cue from them and ignores the saints.

These days, indifference to religion has assumed gigantic proportions. Some say it is due to the numerous counterattractions which, in course of time, dissipate man's mental and emotional energies. But if men really believed the exciting and dramatic facts of their faith, films and football would soon cease to be drugs. Some say the spiritual slump is due to the grim uncertainty of the times. But over against this, what Voltaire said long ago is very true, that calamities and catastrophies drive people more effectively to religion than anything else. The

100

real explanation is that we are now reaping the aftermath of the spiritual sabotage staged by the Victorian intellectuals. What then was confined to the elite has now percolated down to the masses. Talk to any man you like in a factory in one of our industrial cities, and you will be told that scientific knowledge has utterly disproved and discarded religion.

It is salutary to remember that, even in the first century, Christianity was faced with a clever and corroding skepticism, and so we find Peter urging Christians to give a reason for the hope that is in them. By this, the apostle did not mean that you can argue anyone into the kingdom of heaven. If that reminder was necessary in the first century of the Christian Era, it is absolutely imperative now. We must never allow ourselves to be mesmerized by scientists who tell us that they deal with fact while we dabble with fiction. We must face them not only with the meekness Peter enjoins but with the blazing confidence which knows our religion is not something up in the clouds, but something very much down to earth—not nebulous theory but solid fact. Supposing tonight you found yourself up against Fred Hoyle, defending the Christian faith, how would you proceed to give a reason for the hope that is in you?

For myself, I would point him to the *fact of Christ*. At the beginning of this century there were a few odd scholars who questioned whether Jesus ever lived at all. That school of thought is now utterly discredited. We may doubt the existence of Julius Caesar, Henry VIII, Napoleon Bonaparte, but never that of Jesus of Nazareth. The last nineteen centuries of human history are quite inexplicable without him. However much he puzzles us, and in whatever category we finally place him, there is one thing to which we must all agree—the impact of his personality upon mankind has been unmistakable. Round about him he gathered a small band of ordinary men. From infancy they had been reared in the strictest orthodoxy in the Jewish faith, taught to believe there was but one God—Jehovah, one exclusive means of salvation—the law. For three short years they watched him as he toiled and prayed, wept and laughed, blamed and blessed; and so overwhelming was the impact he made on them that suddenly and dramatically they jettisoned the faith of their fathers and said, "Here is God himself walking among us." Subsequently they proved that this was no passing fancy by going out against fearful odds to suffer, to bleed, and to die for him. Ah, yes, murmurs the cynic. Jesus lived in pre-scientific days, when it was easy to impress the ignorant, but if he were to come back today? To that there is only one answer: He doesn't need to. He still has the power to make men do the maddest and most impossible things. Within living memory, he made James Chalmers leave Cambridge and risk his life among the cannibals of New Guinea. When Chalmers came back and thrilled vast audiences with tales of his adventures, his friends begged him to stay at home—but he would not hear of it. Back he went to New Guinea to martyrdom at the hands of the cannibals he sought to save. Mad, no doubt, but Jesus can still inspire his followers with a divine and uncalculating madness. If you judge Albert Schweitzer by worldly standards, the man is crazy. Why an eminent scholar, a brilliant musician, the principal of an ancient college, should casually brush it all to one side and busy himself for the rest of his life in the swamps of French Equatorial Africa, I cannot for the life of me understand. It violates all the canons of reason and common sense. There is but one explanation. Christ is not a spent force in history. By the power of his personality he still attracts not only the ignorant but the learned and illustrious, and sends them out to the ends of the earth to do the maddest and most impossible things in his name.

Returning to Mr. Hoyle, I would point him to the *fact of the Church*. From the very beginning, the logic of things has been piled high against the continued existence of the Christian church. If you had been there that day they hauled a dead

Christ down from the cross, buried him in a deep tomb, and rolled a heavy stone right over the top, you would have given up every vestige of hope. With Pilate, and Caiaphas, and Herod, you would have gone home believing that that was the end. The Christian church is dead and buried beyond the faintest hope of resuscitation or resurrection.

Strangely enough, it has always looked like that. Down the centuries the gravediggers have always gathered around her tomb. Our own brilliant Scottish philosopher, David Hume, tried to bury the Church in the tomb of a cool, philosophic skepticism. Voltaire, his contemporary, confidently asserted that she had one foot in the grave and nothing on earth could save her from death. Yet, in spite of all those gloomy prognostications, somehow or other, the Church has managed to survive. The gravediggers have not yet dispersed. They are waiting to lay her to rest in the tomb of politics and economics, but they are up against an incalculable element which will thwart them in the end. The Church may go down, but, like the eternal phoenix, she rises from the ashes of her own defeat. How often men have buried her, and how often she has burst the fetters of the tomb and risen with new power. How often has she confounded the wise and played havoc with their confident predictions. Near the end of World War II, a German scientist explained to me a new incendiary bomb which could not be put out. From the water used to extinguish it, it borrowed oxygen which made it burn all the more fiercely. How foolish men have been to try and extinguish the divine Light. The bloodier the persecution, the fiercer the opposition, the more concentrated the attacks, the brighter the flame of the Spirit burns. "This is the Lord's doing; and it is marvellous in our eyes."

There is one crucial fact which the gravediggers always forget—the church of Christ on earth is not a human but a divine institution. Therefore nothing that evil man can do will ever finally crush her. Belittle her if you will, sneer at her ambassadors and representatives, call her weak and worldly and corrupt, but remember this: she has weathered the storms and cataclysms of the centuries, and in the end she will conquer. In our despondent moods we would do well to remember the words that come to us across the chasm of the centuries like the blast of a mighty trumpet: "And upon his rock I will build my church; and the gates of hell shall not prevail against it."

And finally, I think I would point my atheist to the *fact of Christian experience.* It is possible to believe in God the Father Almighty. It is possible to believe in Christ as a supernatural being. It is possible to believe in the Church as a divine institution—yet lack all feeling of conviction within our hearts. It is only in the blazing reality of Christian experience that these tremendous facts come to life and speak to us with new meaning. Fred Hoyle talks of Christian belief as an escape mechanism. Where does material come from, he asks, and answers his own question: material simply appears, it is created. "This may seem a very strange idea," he adds, "but in science it does not matter how strange an idea may seem, as long as it works."

If you like, the idea of Christian salvation is strange: strange, surely, that God Almighty should bother himself with our little problems here on earth; strange that he should become incarnate in man to redeem broken and bruised humanity; stranger still that he should transform us by the same power to become the men and women we would like to be. Yes, it is strange—"but it works." To call the fact an escape mechanism is to be crassly unscientific. You can take it and look at it through your long-distance telescope; you can peer at it through your microscope; you can weigh it carefully and sift it in your scientific balances; you can subject it to the severest tests known to man—and one solid indissoluble fact emerges—"it works."

Henry Drummond, who made such an impact on the religious life of Scotland at the end of the last century, had his critics. Some dismissed him as a charlatan—a man who achieved distinction not through merit but through a happy combination of fortuitous circumstances: a well-to-do home, a handsome appearance, a facile pen, and a brilliant platform manner.

These, not character, explain his influence, they said. Then came tragedy. Drummond was suddenly smitten by a mysterious disease of the bones, and he lingered on for two years in excruciating agony. The dying man continued to exude the same boyish, indomitable optimism, the same unruffled equanimity, the same infectious buoyancy of spirit. The critics were confounded. Yes! this thing we call religious experience is quite inexplicable—but "it works."

CHAPTER SIXTEEN

Is Christianity Out of Date?

> *"I tell you truly, till heaven and earth pass away, not an iota, not a comma, will pass from the Law until it is all in force."*
> —MATT. 5:18 (Moffatt)

IN GLASGOW SOME YEARS AGO A DEPUTATION OF MINISTERS met certain civic representatives to negotiate for a church site in a new housing area. After a long discussion, one of the magistrates present said: "If we were realists we would refuse this site, for religion is dying and within one generation the Church will have disappeared altogether." The man, no doubt, was ignorant and bombastic, but he was speaking for millions in our world, who take it for granted that the church of Christ is finished.

We meet the same attitude among many of the intellectuals. Sartre, the philosopher, has called one of his books *The Age of Reason*, implying that the Age of Faith is dead. Faith, he would concede, was a useful enough crutch to lean on when man was bowed down under the heavy load of superstition, but now that enlightenment has come, we can throw away this supporting prop and walk on our own feet.

The vast majority of people, however, would agree neither with the Glasgow magistrate nor the French philosopher. They may not be ardent believers or sworn disciples, but they would hate to see religion go. They believe it has had a stabilizing effect upon the community. From bitter experience they have learned that, wherever it has disappeared, all sorts of sinister substitutes and monstrous evils have rushed into the accompanying vacuum. With genuine concern they have stood by and watched it abdicate its sovereign rights, in province after province of society—in art, education, politics, and economics —till all that is now left to it is the narrow, slippery precipice of individual experience.

We could meet this prevailing pessimism by pointing to the many signs of hope that have begun to appear. The traditional gulf between science and religion, if not entirely bridged, is considerably narrowed. There is a curious return to religion among the intellectuals—the poets, philosophers, physicists, and the like—and in the universities it is stronger than it has been for many generations. In the United States of America there has been a phenomenal increase in church membership and something akin to a religious revival over the last few years.

All that may be true, but if we are to regain our confidence we must probe deeper. Is it not possible that what has already happened in Europe will in time engulf America too? On what ground can we build our hopes for a coming revival? What reason have we for supposing that religion is indestructible and can never become obsolete?

The answer is that we have Christ's own word for it. Did he not say, "I assure you that, while heaven and earth last, the Law will not lose a single dot or comma until its purpose is complete"? Paganism may abound, materialism may appear strong and triumphant, the masses may be indifferent, but according to Jesus there exists in the depths of human nature certain basic needs which only religion can satisfy. These needs remain constant from age to age. The truth is that we have been made in the image of God, and as we confront life in all its maddening complexity, we make certain demands of it.

WE DEMAND COHERENCE

A little boy taking a watch to pieces to see how it works, Isaac Newton seeing an apple drop and deducing the law of gravity, Archibald Fleming peering at blue-mold on a slide and discovering penicillin—all have one thing in common: an insatiable curiosity, a thirst for explanation, a demand for a principle of coherence behind the puzzling phenomena of life.

No thinking being can look out on this world without asking questions. Think of the immensity of the universe, in which this earth is but a "minor and discreditable planet"; to say that modern science has dispelled the mystery and explained it adequately is the height of lunacy. The world, before Copernicus, was mysterious enough, but now since giant telescopes have scanned the heavens, it has become a thousand times more mysterious. The countless universes, the innumerable planets, the unfathomable distances, the frozen stars, the flaming suns speeding through empty space, are facts which deepen, rather than diminish, the mystery—facts before which the reason staggers and the imagination boggles.

Think of the dark inscrutability of nature—so beautiful and alluring on the one hand, so cruel and ruthless on the other. Blake saw nature as a transparent screen revealing the Eternal, but Huxley saw it "red in tooth and claw."

> Tiger! Tiger! burning bright
> In the forests of the night,
> What immortal hand or eye
> Could frame thy fearful symmetry?
>
> . . . .
>
> Did He smile His work to see?
> Did He who made the Lamb make thee? [1]

Think of the greatest mystery—personality itself.

Why were men not made in one mold and shaped to a single pattern? Why is one man a Himmler, murdering millions of his fellow mortals, and another man a St. Francis, loving lepers better than life itself? There is one thing more marvelous than the marvels of modern astronomy, the astronomer who makes the discoveries.

[1] William Blake.

Many slick theories have been advanced to explain the mysteries. The materialist would have us believe that these things just happened—that the universe and life and Shakespeare and ourselves are the products of a chance collocation of atoms, the result of a gigantic stroke of luck, which took place in the dim past. I submit that, in our sanest moments, we refuse to accept such an explanation. We demand a principle of coherence behind it all.

Therein lies the strength of the Christian religion. It claims that behind the towering mysteries, the baffling contradictions, the puzzling phenomena, there stands a God who is also "the Father Almighty, maker of heaven and earth." A claim so shatteringly simple that even a child, long before he has mastered the alphabet, can grasp it. As long as mystery remains, religion will continue to appeal, for it offers the mind what it desperately longs for—a principle of coherence.

We Demand Comfort

"If I had my ministry to live over again," said Joseph Parker, "I would more frequently strike the note of comfort." The cynic immediately pounces. "Ah," he says, "there you go again at the old game of making religion an opiate, a drug, a chloroform mask, in which the weak stick their faces, because they cannot stand up to the pain of life." It is true that certain expressions of religion may degenerate into that sort of thing, but comfort in the real religious sense comes from the Latin root meaning strength, and far from encouraging men to escape, it encourages them to face life with steady eyes and to endure as seeing Him who is invisible.

You have only to take the great heroes of the faith to see how gloriously true this has been. Paul waxed lyrical about comfort ". . . the God of all comfort; who comforteth us in all our tribulation, that we may be able to comfort them which are in trouble, by the comfort wherewith we ourselves are comforted of God." Yet to him, it was never a drug sapping his will, but a source of strength sending him out against staggering

109

odds, to suffer and to die for the faith he so boldly proclaimed. Edward Wilson, in his letters to his wife, constantly speaks of the comfort he derived from religion, but only a fool would seriously claim that it made life easier for him. According to the testimonies of his companions, on his last fateful journey to the Antarctic with Scott, he was a tower of strength to whom all instinctively turned when crisis came and things went desperately wrong.

The modern pose of strength and self-sufficiency is the shallowest of all myths. The contemporary cults, the fashionable ideologies, the ersatz gods may condition our thinking and appeal for a season; but they are not profound enough to alter the deep basic desires of the human heart. Communism sneers at religion as the opiate of the people, but during the last war, when the Russians suffered enormous casualties, Stalin was wise enough to modify his policies and open up the churches. Faced with the demands of suffering and death, he knew that the comfort of faith made a tremendous difference to national morale.

A novelist like Arthur Koestler, who knows his world so well, understands how deep and clamorous in human nature is this demand for comfort. It runs like a deep undertone in all his books and cries out in the words and actions of his chief characters. Hydie is but one example: a self-sufficient, sophisticated young woman, who has plenty of money, mixes with the best society, and breaks all the commandments. Outwardly she appears cynical, careless, completely emancipated, but inwardly she feels a desperate longing for someone who really understands her. Sitting in a shining limousine at a fashionable Paris funeral, she feels lost and longs for the comfort of religion. "The place of God in the world had become vacant," said Koestler, "and a wind was blowing through it as in an empty flat before the new tenants had moved in." Yes, as long as heaven and earth last, man's heart must cry out to a God who understands.

WE DEMAND COMPLETION

Coleridge was in some sort of trance when he composed that unusual poem called "Kubla Khan." He was only getting under way when there was a knock at the door. He rose up to answer it, and when he came back the spell was broken. The muse had departed, the inspiration had vanished, and he could not continue. Reading it, we experience a feeling of disappointment. It is obviously a broken-off piece, a mere fragment of the intended whole, and we resent the callous interruption.

But surely the most tragic thing in all the world is to see a life full of promise callously interrupted and brought to a full stop. John Keats in his early twenties, aware of his own genius and knowing he is doomed to die, feels this inner resentment when he cries:

> . . . I have fears that I may cease to be
> Before my pen has gleaned my teeming brain.

Robertson Smith, an outstanding Oriental scholar, died of tuberculosis when he was still in his thirties. It does not help to say that, young as they were, they had done their work and left the riches of their genius to posterity. They could have left much more, if only they had not been so callously interrupted. No wonder Bernard Shaw, rebelling against the allotted span of threescore years and ten, said that man needs at least three hundred years to understand his world and do justice to his accumulated experience.

That is why, in the end, the philosophy of materialism is so bleak and barren. Its horizons are too cramped. Its vision is too narrow. It has no answer for the desperate desire for completion that throbs in the depth of personality. This explains why religion will never be defeated or dethroned, for it sees beyond the varying and the visible, the permanence of these things not seen: beyond the fragmentariness of our mortal existence, the ultimate fulfillment of life. To our demand for completion it provides an answer.

CHAPTER SEVENTEEN

The Meaning of the Sacrament

"What mean ye by this service?"
—Exod. 12:26

FOUR TIMES A YEAR IN THE CHURCH OF SCOTLAND WE SOLEMN-
ly assemble to celebrate the sacrament of the Lord's Supper.
For centuries our fathers have done it before us, and our children will continue the practice long after we are gone. The
name given to it has varied from age to age and from culture
to culture. Some refer to it as the Mass, others as the Eucharist;
we simply call it Communion.

Millions of our contemporaries dismiss it as a silly superstition—a shunning of the grim, inescapable challenges of life,
a cowardly escape from reality. It is a symbol, they assert, not of
man's idea of the Holy but of the pathetic primitivism into
which he so frequently lapses.

Even many of those who scrupulously observe it remain
blind to its inner significance. It is for them a hazy hallowed
substitute for the magic and mumbo jumbo of our remote ancestors. It exercises the same strange compulsion as touching
wood, throwing salt over the shoulder, or skirting a standing
ladder. It is a cheap insurance against life's sudden and unexpected calamities.

But to us, what does it convey as we stand on the threshold
of an atomic age—looking out on our complex and chaotic
world, listening to the rumbling volcano of our times, and waiting tensely for its eruption? What precisely are we getting at as
we pass around the bread and wine, symbols of a body broken
and buried some two thousand years ago? Has it any relevance?
Does it make sense? What meaneth this service?

COMMEMORATION

Far from being an escape mechanism, it is a handling of the solid, tangible stuff of history itself. The Greeks had a pantheon of gods—Zeus, Aphrodite, Hermes and many more—who dwelt amid fleecy clouds on Mount Olympus and were never seen on the solid earth. Very different is the God Jesus, who suffered under Pontius Pilate and died when Caesar Tiberius reigned in Rome. Here we are dealing not with insubstantial myths but with facts every bit as hard and sharp as the nails that transfixed Christ to the cross.

In Britain's history certain events stand out: the defeat of the Spanish Armada, the victories of Trafalgar, Waterloo, and the Battle of Britain. We cannot forget them because they remind us of a danger that was overwhelming and of a deliverance bordering on the miraculous. These, momentous as they are, pale into insignificance before the event we commemorate through the Communion—the death of Jesus on Calvary. This fact has embedded itself in men's memories and has become irrevocably mixed up with their thinking. Empires have flourished and failed, philosophies have appeared and vanished, fashions have come and gone, but this ancient rite "endureth for ever."

In attending a communion service we are doing more than remembering an event shrouded for us in the mists of the past. We are saluting a Person. "This do in remembrance of me," said Jesus, therein lifting his religion out of vague abstraction and grounding it forever in a personality. It is inconceivable that we should assemble to honor Julius Caesar, Frederick the Great, or Napoleon Bonaparte. But it seems right and proper that we should gather to pay homage to Christ. We have come to commemorate not a dead fact but a living person.

COMMUNION

If we let our minds go back to the first supper in the Upper Room, we see before our eyes a miracle in the making. The little band you see there, torn by clashing temperaments, rival ambitions, and disruptive jealousies, became in time the most closely knit community the world has ever known. The solidarity of their witness has no parallel in history. And their secret? They were in communion with Christ.

By far the most vexing question of our time is that of communion. It confronts us in this era of social revolution through which we are passing. How are men to rise above class and party prejudice and work together for unity? It is the dilemma of Africa, where blacks and whites resort to segregation and terrorism. It is the problem of the whole world artificially divided into East and West, making peace impossible.

The proffered panaceas have all failed. Burns's humanistic dream has not come true:

> For a' that, and a' that,
>   It's coming yet, for a' that,—
> That Man to Man, the warld o'er,
>   Shall brothers be for a' that!

Communism, despite its glowing promises, has proved conclusively that it has no answer. A system that resorts to bloody purges and large-scale liquidations in order to silence criticism has not much to offer us in the way of brotherhood.

The secret is to be found in Christian communion. From the very first it was able to break down the estranging barriers of class, creed, and color and to make men one in Christ Jesus. Lloyd Douglas may be a sentimentalist, but when in *The Robe* he makes Demetrius the slave and Marcellus the blue blooded aristocrat sit side by side and drink from the same cup, he is putting his finger on a fact that shook the ancient world to its deepest foundations. Christian communion scorns vague idealism and utopianism. It brings us back to the solid earth with an uncomfortable jolt, reminding us with the aid of visible symbols that human brotherhood is impossible and inconceivable once you divorce it from a belief in the divine fatherhood.

COMMITMENT

Who were those who partook of the first Christian sacrament in Jerusalem long ago? Not a group of philosophers who had met to ponder the problem of existence. Not theologians who had come to give Christ his proper niche in the scheme of things. Not scholars who had gathered to discuss the mystery of the sacred feast. They were men who had burned their boats.

There is a scene in Charles Morgan's novel, *The River Line,* where an R.A.F. officer whose plane had been shot down is hiding in the hills above a French village. He is waiting for night to fall so that he may enter the hamlet, surreptitiously creep through the silent street, and knock furtively at one of the doors. The resulting drama he sees vividly as with a flash of clairvoyance—the guarded knock, the throbbing silence, the muffled footsteps slowly approaching, the cautious half-opening of the door, the frightened female face, the sudden catch of breath as his uniform is recognized, the tearing agony of indecision. What will this woman do? If she bangs the door in his face she ranges herself with the Gestapo. If she takes him in and hides him she courts death for herself and her family and massacre for the entire village. She has only a few seconds to decide and she must commit herself one way or the other.

That picture gets nearer the central core of religion than dozens of books written on theology. It dramatically reminds us that Christian discipleship is never a matter of speculation or scholarship or painstaking research after truth. It is an act of costly commitment in which the whole personality is involved. To sit with Christ around his table is to shout, "Here stand I and I can do no other."

CONSECRATION

This word, though readily understood by our fathers, has become debased and rarely rings a bell nowadays. It is best translated by the word "discipline," which, however unpalatable it may sound, is at least intelligible to the modern mind.

The most moving scene in Monsarrat's novel *The Cruel Sea* is the one in which the corvette is torpedoed. The vessel is sinking fast, but Wainwright, an ordinary sailor, works quietly, removing fuses from the depth charges to prevent their going off at the point of submerging. He feels the stern steadily lifting as the corvette poises for the last fatal dive, but methodically he works on—"unscrew, pull, throw away, unscrew, pull, throw away"—till the job is finished and he dies a hero's death as if it was all in the day's work.

If the Royal Navy's training can do that for a man, how much more the discipline of faith? We are engaged in a war of savage ferocity where we fight not against flesh and blood but against principalities and powers, spiritual wickedness in high places. If we are to come within seeing distance of victory we must produce men and women with the discipline of Christ ingrained in the inner fiber of their being. Did Jesus not once take a bunch of men, less strong and steady than seaman Wainwright by far, and so mold them that they marched out against savage opposition to place the cross of Christ at the center of the world's mightiest empire?

There are over a million members in the Church of Scotland. If we were a disciplined, determined fighting army, what possible combination of forces in this country could stand against us? There are over eighteen hundred names on the roll of my church, people who have pledged themselves to Christ's service and promised to fight for his kingdom. Given the vision and the discipline of faith, we could take Edinburgh by storm.

Let the communion service be one of consecration where, as soldiers of the cross, we accept its iron discipline. May the bread and wine, symbols of Christ's agony and death, remind us of our promises. In the act of partaking, may we pledge ourselves to a more disciplined church attendance—to a greater measure of discipline in our giving, to a fuller and more costly surrender to his will. Then before the throne, at the final investiture, we shall hear honorable mention of our name. Covered with the scars of conflict, we shall hear ourselves addressed: "Well done, thou good and faithful servant: . . . enter thou into the joy of thy lord."

CHAPTER EIGHTEEN

The Child or the Superman?

*"Verily I say unto you, whoso-
ever shall not receive the kingdom
of God as a little child shall in no
wise enter therein."*—Luke 18:17

EVERY CULTURE CAN BE FAIRLY ADEQUATELY JUDGED BY the type of man it holds up for its ideal. The ideal of the Greeks was the perfectly proportional personality, always aiming at the happy medium. The ideal of the Romans was the soldier, embodying the stoic virtues of discipline and courage. The ideal of modern Germany was the superman, ruthlessly trampling the lesser breeds of Europe under his feet. The ideal of the West is the man who gets on—Ramsay MacDonald rising from proletariat to prime minister, or an Abraham Lincoln coming from log cabin to White House. We Anglo-Saxons worship at the shrine of what Henry James so aptly calls "The bitch goddess, success." But the ideal Jesus held up for all the world to emulate was not the perfectly developed personality, nor the indomitable stoic, nor the man of scintillating success. He picked a child up in his arms, raised him aloft for all to see, and said, "This is my ideal for humanity."

Our first reaction is one of surprise. We are inclined to say, "This adoration of children is all very well, but a world at the mercy of their unpredictable whims and erratic ways would be even more chaotic than our own; and putting all sentiment aside, is it not a fact that children are notoriously cruel and heartless?" Why then did Jesus hold up the child as the ideal pattern of behavior? I suggest that this act shows his profound insight and startling originality. Underneath their tearing de-

structiveness and selfish acquisitiveness he detected qualities,
without which no one can be really great or truly religious.
What are these qualities?

A Capacity for Wonder

A child looks out upon the world with eyes full of wonder.
His little soul is seething with curiosity, and throbbing within
him is the divine mystery of things. The springs of wonder and
religion lie close together, and the atrophy of this sense is one
of the main causes of modern irreligion. Under the impact
of a bogus rationalism we have become incredibly blasé. We
have explained away everything in terms of cause and effect,
and we have banished God out of his own universe. No wonder
the poet protests:

> God is a proposition.
> And we that prove him are his priests, his chosen.
> From bare hypothesis
> Of strata and wind, of stars and tides, watch me
>   construct his universe,
> A working model of my majestic notions,
> A sum done in the head.
> Last week I measured the light, his little finger;
> The rest is a matter of time.[1]

Before life's towering mysteries we sit down unmoved. The
miracles of recent discovery we take callously for granted. We
have lost the inexpressible sense of wonder which meets us in
the New Testament. These men could never get used to the
incredible news of the gospel. Christ had come! He had died
for their sins! He had risen from the tomb and defeated death
and evil forever! Before facts so staggering and momentous
they could only stammer out, "This is the Lord's doing, and
it is marvellous in our eyes." "By wonder we are saved," so
wrote Plato long ago, meaning that it takes the shock of a

[1] Copyright 1940 by C. Day Lewis. Reprinted by permission of Harold Matson Company.

stunning amazement to convert a man. It is in this sense that
Jesus is the only Saviour of the world. He shatters our blasé
complacency. He stabs our slumbering soul awake, and he
makes us look out with a child's eyes into the mysteries of
God's world.

> He wakes desires you never may forget,
> He shows you stars you never saw before,
> He makes you share with Him for evermore
> The burden of the world's divine regret.
> How wise you were to open not! and yet
> How poor if you should turn Him from the door!

A Capacity for Belief

The child finds it easy and natural to believe. Skepticism
is foreign to his nature. Modern man is the very antithesis. His
mind has become curiously inhospitable to religious faith. His
spiritual muscles have atrophied. His power to believe in God
has all but vanished. It is not that he is fundamentally irre-
ligious. There are times when, wistfully and desperately, he
wants to believe but finds he cannot. We meet this undercurrent
of wistfulness in a confession made by Charles Darwin toward
the end of his life. He admitted that preoccupation with scientific studies had killed his taste for poetry and music and the
things of the spirit. Darwin's descendants and disciples could
go further and admit that their noisy preoccupation with things
visible has destroyed their sense of God himself. Wordsworth
saw this coming and protested hotly:

> Great God! I'd rather be
> A Pagan suckled in a creed outworn;
> So might I, standing on this pleasant lea,
> Have glimpses that would make me less forlorn.

What has happened? Is it, as Freud suggests, that religion is
a form of childish credulity which the mature personality seeks
to shed? Is it, as Sartre says, that we are now living in the age of

reason, and that faith is a thing of the past? Is it, as so many
today claim, that man is really emancipated and no longer needs
this crutch to support him as he staggers under the burdens of
life? Far from it. The reason is that the springs of religion deep
within our nature have been stopped, as the wells of Abraham
were stopped by the Philistines in the valley of Gerar long ago.
Jesus, in holding up a child before our eyes, is saying, "You
were created for belief in God. You are incurably religious.
In the secret depths of your personality there is an eternal
kinship between you and your Creator." May God give us
grace to shed the pseudointellectual and cultural dross that
chokes our souls; that the springs of faith within may once
again flow with their pristine force.

A Capacity for Sheer Natural Simplicity

Who can resist the God-given naturalness of a child? Which
parent has not been embarrassed by the untimely outspoken-
ness of his carefully tutored offspring? Karl Barth, the famous
theologian, tells a delightful story against himself. There was
a time when he decided to dispense with grace before meals,
thinking it had become a mere formality. One day, while
entertaining a well-known Scottish theologian at lunch, he said
grace, whereupon his little son piped up, "Daddy, why do you
only pray when strangers are in the house?" Under the impact
of a synthetic and soul-destroying culture we have lost that
divine spontaneity. We have become horribly complex and
sophisticated. We have raised barricades and bulwarks of de-
fense in the outer fringes of personality and pushed our real
self far into the background. Jesus hated a clever, sophisticated
insincerity. It was the one barrier God himself could not
penetrate. Not even God Almighty can save a man who is forever posing and pretending. That is why men like Voltaire,
Byron, and George Bernard Shaw were incapable of a genuine
religious experience. They were too clever—posers who choked
the deep springs of faith and smothered the still, small voice
of God by their egotism. If I were really up against it, if I

were facing some titanic crisis in my life, if I were really at my wits' end wondering where to turn, I would go not to the clever or the cultured or the successful, but to some simple soul who had the stamp of Christ's character indelibly imprinted upon him.

As an undergraduate at St. Andrews University I witnessed a very dramatic incident. Professor W. P. Paterson was addressing a large student gathering. For over two hours he answered question after question in profound and brilliant manner. Then suddenly he said, "Let us pray," and like a little child dropped to his knees before us all. The effect created was overwhelming. Afterwards a vociferous agnostic remarked, "I found his answers brilliant yet unconvincing, but that last act of his—it got me." The challenge of Christmas is one of simplicity and sincerity. As we go to Bethlehem and gaze at the child Christ, wrapped in swaddling clothes in a common stable, may these words speak compellingly to our hearts. "Verily I say unto you, Whosoever shall not receive the kingdom of God as a little child, he shall not enter therein."

CHAPTER NINETEEN

A Religion That Hurts

"Come down from the cross."
—MATT. 27:42

OVER THE DROOPING FIGURE OF THE DYING CHRIST WAS HUNG the superscription, written in Latin, Greek, and Hebrew, "This is . . . the king of the Jews." The scribes and Pharisees, the soldiers and passers-by, even the thieves crucified on either side of him, mocked and derided. "If thou be the Son of God," they cried, "come down from the cross and we will believe."

That motley throng around Calvary disagreed about everything else in life, but on one thing they were unanimous: this man was no king. If the mark of kingship was power—and it was to all Orientals—this expiring prophet with his head hanging down upon his chest was its very antithesis. He may have performed miracles and saved others, but when the acid test came it found him out. All he could do was to hang there, a public spectacle, the butt of a hate-intoxicated mob.

Yet the demand to come down from the cross was not entirely sarcastic. There was in it an element of wistful expectation. They almost wished he would.

And we are still taunting Christ and coercing him to come down to our own level. Confronted with mysteries that haunt and torment us, we cannot understand why Christ is so annoyingly inactive. "Why is he so hidden and inaccessible?" we keep asking. "If he cared enough for the world to die for it, why does he not come out into the open and do something?"

These men would not let Jesus die in peace. As he writhed in agony, they hurled their foulest insults at his head. His intense suffering called forth not their compassion but all the concentrated spleen and spite of their nature. Infuriated by his silence, they cried: "Come down from the cross; give us a sudden spectacular miracle and we will believe." They found the cross meaningless because:

They wanted a religion they could understand. For centuries the Jews were looking for the advent of the Messiah. From infancy they had been taught what exactly to expect. The coming One was to combine within himself the qualities of a shepherd and a conqueror—a shepherd to lead people into paths of righteousness, a conqueror to crush the opposition, to liberate the masses and herald in the dawn of a new and better order.

There can be no doubt that the Jews who first hailed Christ expected him to conform to type. They wanted a social reformer and a political firebrand, and when instead he chose the way of the cross they were stricken and dumbfounded. A Messiah riding in triumph and trampling his enemies underfoot they could understand. A Saviour pleading forgiveness for those who had crucified him they could not even dimly comprehend. Here was something that mocked their dreams and contradicted their fondest expectations.

This problem is still on our hands. We cannot square Christ with our own preconceived notions. He bursts the normal categories of our thinking and breaks out of the theological strait jackets in which we try to encase him. "Come down to our human level," we cry. "Descend into the political arena, condemn Communism, challege capitalism, help to raise the standard of living, and we will believe in thee."

There are others who are quite prepared to accept Christ as long as he does not claim to be anything more than a man of genius. Strip the Galilean of his mythical divinity, dissociate him from anything supernatural, reduce him approximately to the stature of Socrates, Buddha, and Shakespeare, and we are ready to accord him a place among the immortals. Today, as then, people resent a Christ whom they cannot understand.

One function of religion is to remind us that there are certain questions in life utterly beyond our comprehension. A. J. Cronin, in his autobiography, tells the story of the scientist who addressed his boys' club in London. In a frankly atheistic talk, he explained how the pounding prehistoric seas upon the earth's crust had produced a pulsating scum from which life eventually emerged. In the ensuing discussion a nervous lad with a stammer asked, "Sir, you have explained how the b-big waves b-beat upon the shore, b-but how did they get there in the first place?" There was no answer. The elaborate logic of the test-tube realist crumbled before a question so shatteringly fundamental. And the Cross will never let us forget that, at the heart of religion, here are ultimate mysteries before which human reason staggers and logic is stricken dumb. We can readily see why, from the beginning, it was to the Greeks foolishness and to the Jews a stumbling block. It still is to all who regard the intellect as the final and absolute authority. The paradox of Calvary can only be grasped by faith.

They wanted a religion without costly challenge. On this the New Testament is quite brutally frank. As long as Jesus wrought mighty miracles and impressed people by his unique powers, he had a following. But no sooner did he begin to speak of the Cross and the certainty of suffering than the adoring crowds disappeared into thin air. If the new faith demanded pain and self-denial, they would have none of it. "If you come down from your cross and give us a religion that does not make such impossible demands, we will believe."

But that is what Christ has persistently refused to do. To every demand for an easygoing, accommodating faith he has remained silent, arms outspread on the cross. In Aldous Huxley's book *Brave New World* the unpleasant corners of life are smoothly filed away and pain itself is eliminated by applied science. "What I offer you is Christianity without tears," proudly shouts the controller. And that has been the weakness of the Church now for many, many decades. It has been offering the world a religion with no Cross at its center—Christianity without tears. And we are surprised that it does not save or heal or comfort a living soul.

It seems to me that we are spending a disproportionate amount of our time on diagnosing the spiritual malaise of our age. The techniques and approaches and methods which we hear of now to the point of weariness tend sometimes only to obscure the real issue. Our most permanent need is not the discovery of this device or the employment of that method but men and women who are willing to get hurt following Christ. Such people are ready to resist the pressure of class and creed and conditioning process, and are prepared to commit themselves utterly, till they feel in the most sensitive fiber of their being the terrible anguish of the cross.

Recently, in a theological magazine there appeared a scathing article on Albert Schweitzer, the great missionary. The author attacked his theology as dated and obsolete and the man himself as muddle-headed, misguided, and off-center concerning the fundamental teaching of the New Testament. Toward the end of the article, however, he paused to say:

Perhaps we should not judge Schweitzer by his words but by his deeds; not by his books but by his Christian discipleship; not by his theological conception but by the fact that he took up Christ's Cross and, carrying it to Africa, has not yet laid it down.

That, in the end, is the only convincing argument. The only religion that heals is the one that hurts.

They wanted a religion that offered quick results. "Give us one of your famous miracles," jeered the spectators. "You opened the eyes of blind Bartimaeus, you cured the epileptic boy, you raised Lazarus from the dead; you saved others, surely you can save yourself." But there was no response—nothing save that awesome silence.

The centuries have rolled on, but we are still pressing Christ to produce spectacular miracles. We are still holding him up to ransom and saying in effect: "Dissolve this doubt, remove this mountain, grant me this ambition, and I will fit you into my philosophy of life." This reduces religion to a vulgar utilitarian level, to mere materialism thinly disguised—and incidentally it explains why Jesus in the New Testament never gave way to the popular clamor for the spectacular. He knew that any faith reared on such a foundation was shoddy and produced a character equally shoddy. I dare say Jesus could have come down. The Christ who walked on the heaving sea and who fed the hungry multitude could have done something startling, but he remained where he was. This was the greater miracle. "They would have believed if He had come down," said General Booth. "We believe because He stayed up."

The truth is that Christian experience never conforms to this clamorous demand for spectacular miracle. Even those who enjoyed sudden catastrophic conversions found that the real struggle took place afterwards. Augustine heard the liberating voice in the orchard and the inner storm of doubt was laid to rest forever—but the battle against his powerful passions was a long and bitter one. Tolstoy in the wood stumbled upon the truth in an instant, but according to his own *Confession* it took him years to master his strong appetites. When a young minister fresh from the inspiration of the Keswick Convention bubbled over to Alexander Whyte, the old warrior shook his head and said, "Aye, man, it's a sair fecht to the end."

Paul, who ought to know, assures us that discipleship means crucifixion, and crucifixion in those days was a long-drawn-out, painful process. But the New Testament never stops there. The Christ who died was also the same Christ who rose again to reign in triumph, and in Christian experience we must die to self before the new life surges up within us. There is profound and paradoxical insight in the testimony, "I am crucified with Christ: nevertheless I live; . . . and the life which I now live in the flesh I live by the faith of the Son of God, who loved me, and gave himself for me."

CHAPTER TWENTY

The Art of Passing the Buck

"Thou art the man."
—II Sam. 12:7

IT IS TOLD OF NAPOLEON THAT HE HAD SOMETHING AMOUNT-ing to genius for fooling himself. Faced with disaster, he used to count up regiments that did not exist even on paper. When his staff remonstrated and pointed to the folly of such a habit, he would turn around sharply and exclaim, "Would you rob me of my peace of mind?"

King David of Israel, another man of might, was guilty of something similar. He tried to shut up his religion and his conduct into two separate, watertight compartments and fool-ishly thought he could retain his peace of mind. In a moment of weakness, he sent Uriah the Hittite to his death in battle, in order to possess his wife. You would think that the man who composed the twenty-third psalm would have known better. But he rationalized the whole thing away. "I am a king with certain privileges denied to the ordinary run of mortals. Uriah is a mere Hittite, a base and inferior fellow, and his wife is not in love with him; in fact, I am paying her a great compliment." So he excused himself. But David did not get away with it. The prophet of God painted for him a vivid parable of a sordid trick played on an innocent man. David listened with mounting anger. Finally he burst out with demands for the identity of the miscreant. Nathan looked him straight in the eye and said, "Thou art the man." David might have succeeded in fooling others and himself, but he could not fool God.

We are all past masters at shutting up our ideals and our

conduct in sealed, separate compartments, and of trying to evade personal responsibility for our mistakes and our sorry failures. In politics no party will ever confess to muddle or mismanagement. It is always the rival party which ruins the country and is responsible for its supposed decline. What a pleasing shock we would get if, some day, a leading politician were to stand up and confess, "I have made a ghastly mess of things. I have misunderstood the trends of the times. I have wrongly diagnosed the maladies of my age. I have been proffering the worst possible panaceas. Let me depart, that a better and wiser man may take my place."

In Fitzroy Maclean's book *Eastern Approaches* there is a tragicomic scene of a people's court in Moscow: comic in the sense that the trumped-up charges were so absurd and impossible—smashing a few million eggs, mixing up powdered glass with butter, throwing sand into locomotive engines; tragic because those standing in the dock were already doomed. What happened obviously was that doctrinaire planning had failed to produce the goods—the engines, the eggs, and the butter—and suitable scapegoats had to be found to fool the people.

But this slick solution of life's vexing problems, and this attempt to fool others and ourselves, are not just confined to ancient Semitic kings or modern Communists. Here we are all involved. We are not guilty of murderous tragedies like that of Uriah, or the liquidation of innocent men to cover up our own mistakes, but we are guilty of morally "passing the buck" and of ascribing the world's miseries to other people's sins. At this point we all stand under judgment. Let me illustrate by referring to two classes of people common to our society.

There is the class who call themselves "believers." They may be regular churchgoers, unfailing in their attendance, Spartan in their spiritual discipline, willing to give a fair proportion of their money to the works of charity and the church at home and abroad. But often these good people, though admirably sincere and stanch, are lamentably limited in their vision.

Every novel experiment, every new approach, every courageous effort to break new ground and reach the unchurched masses meets either with their suspicion or their strong disapproval. The great Alexander Whyte of Free St. George's, Edinburgh, would never stoop to such strange and questionable strategies. He just preached the old, simple gospel. The church was thronged, the people were pleased, and everything in the garden was lovely! Religion to such people is a matter of tradition, or aesthetic enjoyment, or a purely private concern. They draw a sharp line of demarcation between the sacred and the secular, and confine Christ to the pigeonhole of a narrow individual experience. Politics, economics, education, and the crucial concerns of modern society lie beyond the orbit of their gospel. They would be shocked to discover that that is precisely the mentality the modern dictators seek to hammer into the gullible masses. "You can wallow in a private, underdone religion to your heart's content and we have no objection, but once you begin to apply it to the big problems of society, we'll clamp down on you and shut you up in prison." Niemoeller passionately believed in personal religion, as every Christian does, but he saw that any genuine religion has social implications which cannot be ignored, and so when he raised his voice in Germany he was suppressed and silenced. Those of us who bury religion in the tomb of the past, who smother it under the suffocating weight of the *status quo,* who make it a matter of private aesthetic enjoyment, are guilty of divorcing the faith of Christ from life and allowing the richest provinces of human experience to fall into enemy hands. We fail to see that if the claims of Christ are true *anywhere,* they must be true *everywhere.* Such Christians are partly responsible for the chaos and confusion of our world, and they cannot evade the divine finger pointing at them, or the divine voice thundering, "Thou art the man."

There is another kind of Christian who, though he thinks the gospel of materialism is bleak and indescribably stupid, sits very loosely to the Christian church. He is the type of man who eats his Sunday-morning breakfast late, or lolls comfortably in a chair with a Sunday-morning paper open on his lap and listens to the radio service. He persuades himself he is taking part in a corporate act of worship and is doing his bit for Christianity. There are millions of such armchair Christians in Britain who turn around to you blandly and say, "We never go to church. There is no need. We hear excellent sermons on the radio." The religion that crucified Christ on Calvary was meant to be something more demanding than listening passively to Sunday-morning services on the radio. To such Christians—nice people as they are—religion is merely a cult of decency, charming over a cup of afternoon tea, or during one of the parson's innocuous calls, but tragic as a bulwark against the savage forces that are tearing our world apart. They, in a very real sense, are responsible for the marching militant godlessness of our times. If at this point your conscience begins to murmur, that is God speaking to you and saying, "Thou art the man."

There is the class of people who call themselves unbelievers. Among them are out-and-out materialists, who truculently tell us prayer is autosuggestion and religion a prejudice of the middle classes, a chloroform mask, into which the weak stick their faces to hide from themselves the real truth about life. To them matter is basic, the Christian church a vast structure of organized superstition which should be either completely ignored or kindly disposed of. Like Voltaire's contemporaries, they would echo the old refrain, "Get rid of organized religion and Utopia is just around the corner." Well, the Nazis did just that, and the result was the liquidation of four million Jews and a war from whose aftereffects we are still suffering. Materialism as a gospel may sound attractive, but it has no respect for human personality. In subscribing to it, a man is in a sense responsible for the mass murder of innocent people and is adding to the sum total of misery that darkens the earth. "Thou art the man."

More subtle and more dangerous are the skeptics, who impress us by the integrity of their characters. They go to no

131

church, subscribe to no creed, believe in no God, yet they are brave, generous, truthful, and astonishingly attractive. How often have you heard the refrain, "I know so-and-so, who does not believe in God, and yet is the most decent of men." Let us give them full credit for their qualities. Indeed, we cannot help being challenged and humbled by some of them, but there is one gaping chink in their armor—the lack of a positive faith and the consequent inability to pass on to their children any coherent philosophy of life, which they will so desperately need in the stormy world of tomorrow. They may be admirable, but they lack what men of faith have always possessed—an overplus of moral energy which they can pass on to others.

In a powerful French novel called *La Morte* we get a vivid picture of the home of a skeptic called Bernard and his profoundly spiritual wife, Aliette. She takes ill, and he calls in a brilliant doctor named Vallehaut, also a skeptic. He brings with him as a nurse his niece, Sabine. Then the tragic thing happens. Sabine falls in love with Bernard and poisons Aliette. The doctor is perplexed by the unexpected death of his patient. At night, unable to sleep, he stealthily descends the stairway and creeps into the living room. The moonlight is streaming through the window. He stands before a medicine cabinet in the wall, opens the door, and looks within. Sure enough, on the shelf where the poisons are kept, there is an empty space. Under his breath he says, "Aconite, the poison that scarcely leaves a trace." Just at that moment he hears a step in the corridor and he moves back into the shadows. There is a rustle of silk. Sabine enters, carrying a little bottle in her hand. Swiftly the doctor seizes her by the arm. "Defend yourself," he cries. "That was a terrible premeditated crime. Defend yourself! You count on my weakness, but horrible as the duty is, I will hand you over to justice." She looked at him scornfully, her face white in the moonlight, and said, "Don't be in such a hurry. I will tell the jury that I am your apt pupil." "Are you mad?" he asked, "my pupil! You know that I have always lived a noble life." "You surprise me, Uncle," she replied coldly; "you talk about nobility and justice and honour, but you are the man who robbed me of a faith that would enable me to believe in these ideals. You have taught me that there is no home in the universe for right or wrong. You are my teacher. You are my instructor, and I shall tell that to the jury."

Are you proud of your clever skepticism, proud that in spite of it you are able to live a decent and noble life? May I be allowed to put one searching question? Have you an overplus of moral energy which you can pass on, not only to your children, but also to people like Sabine in the novel, who have known the fierce assault of temptation and who are clamoring for some power to deliver them? If by your smooth and slick skepticism you are undermining the ideals of your generation, and are causing your brethren to stumble and fall, you are guilty of an enormous crime. The voice of God singles you out and says, "Thou art the man."

But remember, mercy, not judgment, is the last word. David, who was crushed by the memory of his sin, was the same man who prayed, "Have mercy upon me, O God. . . . Create in me a clean heart . . . ; and renew a right spirit within me. . . . Restore unto me the joy of thy salvation." The God whom Christ revealed is not a sentimental Santa Claus. He is a God who pricks the bubble of our complacency, shatters our good opinion of ourselves, stabs awake our slumbering conscience, and cries to us in the midst of our sins, "Thou art the man!" He is also the God who pities us even as a father pities his children, and restores us to newness of life by the strength and power of his love.

CHAPTER TWENTY-ONE

The Basis of Certainty

*"For I am persuaded, that
neither death, nor life, . . . shall be
able to separate us from the love
of God, which is in Christ Jesus our
Lord."* —ROM. 8:38-39

IN A MODERN NOVEL, "THE AGE OF LONGING," A SOPHISTICATED
American girl falls in love with a Communist. She hates his
politics, detests his philosophy of life, and is repelled by his
untutored, almost savage, manners. But there is one thing about
him that casts a spell over her—his amazing certainty. She
herself has neither faith to inspire her nor creed to fight for. In
the face of the enigmas of life she is confused and bewildered.
By contrast he, though quite flagrantly amoral, has one goal
in life toward which he strives with all his might and main. It
never occurs to him that his entire scheme of things might be
wrong. He is completely certain. Confronted by such confidence,
people unanchored to any conviction find themselves fascinated
and even hypnotized.

That is a dramatic symbol of the dreaded disease which
threatens to rob the world of its health and vigor. Our civilization is in danger of collapse because the cancer of uncertainty is
clawing away at its very vitals. In politics the enigma of Russia
offers a glaring example. What precisely are her aims? A policy
of expansion or consolidation; naked imperialism or internal
development; war or peace? If only we knew for certain, the
millions we spend on armaments could be put to better use.

Science used to be so certain and cocksure of everything
under, and including, the sun. It possessed the key to all secrets

and in its own eyes was the saviour of mankind. That cast-iron confidence has now disappeared, and bewildered voices are heard asking if its accumulated knowledge is to be used merely to blow man and his greatest achievements to hell.

That prevailing sense of confusion meets us at the very core of personality itself. Men and women all around us have not merely surrendered the moral standards which our fathers considered sacred: they have begun to question the rationality of the universe and the very purpose and meaning of life as well. Like Koestler's sophisticated heroine, they may pose and rationalize and argue, but deep down they feel mortally stricken. Like her they are instinctively drawn to those whose feet are planted on solid rock while their own are caught in the sinking sands of circumstance.

Perhaps that explains Paul's abiding influence. It is true that he lived and risked and sacrificed as no one else did, but that alone would not explain his perennial appeal. He towers above the centuries like some great Rock of Gibraltar which cannot be stormed. This man was absolutely convinced of the truth he had found, and nowhere does this burning conviction find more powerful expression than in the memorable words: "For I am persuaded, that neither death, nor life, . . . nor principalities, nor powers, nor things present, nor things to come, . . . shall be able to separate us from the love of God, which is in Christ Jesus our Lord." Across the centuries we can hear the echoes of ringing certainty that was the secret of his immense assurance.

The Certainty of Christ

We live in an age of crisis and confusion. It is possible to believe that our epoch has been cast into one huge melting pot and certainty has disappeared forever. But not for one moment must we imagine that the world of Paul was simple, straightforward, and uncomplicated.

We have centuries of Christian history behind us: Paul started from scratch. There was little in life he could be sure of. Barbarism reigned supreme in the ancient world: the fury of persecution blazed all around him. The future of the Church hung delicately in the balance. Even his stanchest disciples deserted him. In the end he found himself in a death cell in Rome, waiting for the quick knock at the door, the peremptory summons to execution. Throughout he remained certain of one thing—Christ.

Ah yes, you retort, but after all he was a contemporary; we are different, we belong to a different world. How can we ever become certain of a dim shadowy figure shrouded in the mists of antiquity?

> Dim Tracts of time divide
> Those golden days from me.
> Thy voice comes strange o'er years of change,
> How can we follow Thee?
> Comes faint and far Thy voice
> From vales of Galilee;
> Thy vision fades in ancient shades
> How should we follow Thee?

It is precisely at this point that Paul's testimony is conclusive. "Yes, it is true that Jesus and I were contemporaries," he concedes, "but I never knew him in the flesh. It was years after he was crucified, dead, and buried that I met him in broad daylight outside the gates of Damascus. Other experiences in my life are open to question, but this one never. The only thing in the world I am absolutely and utterly sure of is that I saw the risen Christ and heard him addressing me by name: 'Saul, Saul, why persecutest thou me?' "

And that has always been the strange thing about Jesus: he is not a historical figure but a living contemporary. There are tens of thousands living today who would echo Paul's confession. Baffled by mysteries, crushed by contradictions, haunted by cruel enigmas—there is only one thing we are certain of—Christ. On him we are prepared to stake our lives.

THE CERTAINTY OF GOD

Does God care? That is not an academic question posed by theologians in secluded classrooms. It is the most relevant question of our day. This earth of ours is but a minor and discreditable planet in a vast universe. Think of the countless million stars evolving through the countless million years in the silent immensities of space and ask yourself if it is easy to believe God cares. Nature—with its grim inscrutabilities, deluge, earthquake, tornado—returns an equally empty answer. Think of what happened in the Ionian isle of Greece—mountains on fire sliding into the sea, entire towns obliterated, and hundreds of women and children buried under tons of rubble and fallen masonry—and answer me, do you find it easy to believe that God cares? History is supposed to have a divine purpose running right through it, but think of Ivan the Terrible, Napoleon, Stalin, Hitler, and tell me, do you find it easy to believe in a God who not only controls but also cares?

Men have brooded over the problem and given their different answers. Aristotle, some four centuries before Christ, wrestled with it and concluded that God was the "moveless mover" who gave the initial push and then sat back watching the stupid game from a distance. That may be better than atheism, but it is no comfort to a man broken by sorrow and caught in the vortex of circumstance.

Matthew Arnold lost his son, and, sore at heart, he turned to philosophy and poetry, to Spinoza and Goethe and Marcus Aurelius, seeking comfort. He arrived at no passionate conviction that behind the heartbreaks of life there was a God that really cared. His search brought him only a vague consciousness of "something not ourselves that makes for righteousness."

And yet if there is no such conviction integrating the whole of life, it is impossible to be carefree and happy. But if there is, where are we to find it? Not in nature or history, not in philosophy or literature—only in Christ. Paul was a genius, conversant with all the philosophical systems of his day and

137

steeped in literature. When he talks of the love of God, however, he has no recourse to logic or clever dialectic. The one argument he uses is all the more impressive because it is so shatteringly simple. "God," he said, summing it up in one concise sentence, "God commendeth his love toward us, in that, while we were yet sinners, Christ died for us."

THE CERTAINTY OF IMMORTALITY

The desire for immortality has pursued men down the ages, and certain classic arguments have been paraded to carry conviction to the human heart. The great German philosopher Kant argued that there was so much injustice here that there must be an afterlife where the moral balance would be redressed. Personally, I find such logic singularly unconvincing.

There is another argument which holds that the yearning for survival is like a powerful instinct throbbing at the core of man's being. As an answer has been provided for all other human instincts, this one can't possibly be an exception. Again I must confess that it fails to carry conviction to my heart.

Better still is the argument which maintains that personality is by its very essence indestructible. You cannot conceive of the genius of Shakespeare, the mind of Michelangelo, the soul of St. Francis obliterated by a germ or a chance accident. This argument, I admit, makes a tremendous appeal to me, but even it cannot silence the voice of doubt, or quell the fear that mocks my fondest dreams.

In the last resort there is only one argument that silences doubt and carries conviction to the heart of man. It is the one so passionately trumpeted by Paul in this lyrical passage: I am certain of Christ, therefore I am certain of two other things—the character of God and the fact of immortality.

At school the teacher who taught us geometry started off with a certain basic theorem. On the blackboard he would draw a diagram and prove to us that the angles at the base of an isosceles triangle were equal. Once this was accepted, he went on to show how certain corollaries followed from it as a matter of course. Well, Paul demonstrated one basic theorem to the world—the fact that Christ lived and died and rose again. Because this was true, you could never again doubt the Son of God or question the reality of the world to come and life everlasting.

CHAPTER TWENTY-TWO

The Light That Must Not Fail

"Give us of your oil; for our lamps are gone out."—MATT. 25:8

MOST PEOPLE EQUATE CHRISTIANITY WITH GLOOM. THE popular belief that religion makes a man narrow, inhibited, and negative is both curious and challenging. Swinburne may have been thinking of grim-faced Victorian piety when he penned the famous line:

Thou hast conquered, O pale Galilean;
The world has grown grey from Thy breath.

Aldous Huxley, in one of his earlier essays, calls it a case of spiritual gangrene. H. G. Wells is more blunt: he claims that it kills all natural happiness and dries up the very juices of life.

How this common conception has arisen is most puzzling, especially since Jesus, time and again, compares his religion to a marriage feast, where the dominant note is one of joy. His disciples, when they ran into the streets of Jerusalem to proclaim the gospel of the Resurrection, displayed such excitement and exuberance that they were accused of drunkenness. Let a man absorb the spirit of Christ, and one thing is certain: he cannot hide his zest for life nor suppress the joy that spontaneously wells up within him. And, today, no matter how clear our vision or realistic our approach, it is impossible to shatter the crust of public indifference until, as Christians, we have recovered this lost radiance of the faith—the joy which the world can neither give nor take away. This parable of the wedding guests waiting the coming of the bridegroom drives home three main points.

THE NEED FOR VIGILANCE

Throughout the New Testament this note sounds forth like a clarion call. Jesus knew that evil was no vague, dormant principle but something savagely, energetically offensive. It is when Christians, drugged by the soporific climate of public apathy, become careless that evil seizes the chance to gain the mastery.

Luther, in one of his legends, reminds us of this in characteristically robust manner. Satan called together a council of his chiefs to submit their reports.

"I let loose," said one, "the wild beasts of the desert on a caravan of Christians and their bones are now bleached in the sands."

"What of that!" said Satan. "Their souls are all saved."

"I drove the east wind against a shipload of Christians," said another, "and they are all drowned in the sea."

"What of that!" said Satan. "Their souls are all saved."

"For ten years," said another, "I have labored to lull a group of Christians to sleep and at last I have succeeded."

"Then Satan shouted for joy," said Luther, "and all the stars of hell rejoiced together."

"Watch," said Jesus, "I say unto you all, watch. Watch therefore for ye know not the day nor the hour." And today, when evil is better organized than ever before, that demand for a keen-eyed readiness is no less urgent.

"Watch," he says, "in the realm of history." If the Christian church had been alert and vigilant in the past, the forces of evil could not have straddled the world so easily and flaunted their success so arrogantly before our eyes. Materialism would not have assumed its present proportions. A ready, resolute Church would have checkmated it by staking out Christ's counterclaims. The majority of church members still subscribe to the heresy that, for Christians, politics are out of bounds, that

141

they must on no account get mixed up in this dirty game. So, in Germany, they played into the hands of Hitler; in Russia, into the spider web of Stalin; in America, into the short-sighted intrigues of democratic demagogues. Willy-nilly we are being forced to enter the field of politics as the prophets Isaiah and Jeremiah were, as Calvin and John Knox were, as Niemoeller and Bergraev were in our own day. This is Satan's most strategic stronghold. "Watch, therefore," says Jesus.

"Watch," says Jesus, "in the realm of ideas." The coming of the Copernican, Darwinian, and Marxian revolutions found the Church mentally asleep. Far from adapting herself to the new outlook, she resisted bitterly. The Church called in question the solid chunk of truth inherent in them all, and so lost valuable ground which only now are we beginning to regain. A church that sets her face against new ideas is not likely to give much of a lead in this fast-moving revolutionary age. The stresses and tensions of our time are forcing us to see that only a Christian who knows his Bible and is able to relate it to present-day dilemmas has any hope of winning the attention of the masses. Not only must we become soundly indoctrinated in our faith, we must become more articulate. Every Christian must be prepared to give a reason for the hope that is in him. Here, as never before, we must be wide awake and alert.

"Watch," says Jesus, "in the realm of Christian witness." We loudly lament the chilling apathy and corroding secularism of the times, but the truth is that this is a day of unrivaled opportunity, if only we were prepared to seize it. It is common to accuse the clergy of being lazy and out of touch with their generation. Who has not heard their preaching described as prosaic, dull, and pedestrian? There is some truth in the charge, though it can be grossly exaggerated. Some are lazy and some are stupid, but most of them are sincere and hard working. If you ask me wherein lies the basic weakness of the Church of Scotland I would unhesitatingly answer, in the laity. During the war I came across the strange sect we call the Mormons, and though I could not stomach their theology or accept their dogmas, they impressed me more than I can say. They possess no regular clergy, but they make an impact in the life of the United States out of all proportion to their number. And their secret? They are all missionary minded. The banker, the lawyer, the teacher, the doctor, the insurance agent, the journalist—all are live wires, conducting the current of their faith to the world.

And in Scotland we have a system which is the envy of all other churches—the eldership. The ordained elder can enter every house in the congregation assured of a ready welcome. What a God-given chance! What a golden opportunity for bridging the chasm yawning between the Church and the world! What a powerful instrument if we were only prepared to use it in the interests of Christ's kingdom! Don't tell me that this machinery is cumbersome and out of date. It still reaps rich dividends. In any venture, you can predict with something amounting to scientific precision that the district with a good elder will respond, on a spiritual and material plane, much better than the district with an indifferent one. Revival will come, not when the clergy begin to preach better and more interesting sermons, but when lay men and women—elders, deacons, and ordinary members—begin to witness more intelligently and enthusiastically. When, in the words of Holy Writ, they wait expectantly "with their lamps lit" and "their loins girt." This parable also drives home:

THE NEED FOR A PERSONAL CREED

"Give us of your oil," cried the foolish virgins in a panic. "Not so," replied the wise ones, "lest there be not enough to go round." This abrupt refusal may sound selfish and unchristian, but the point this part of the parable makes is that in life, at the midnight hours of challenge and crisis, there are certain commodities you cannot borrow.

It is a mere truism to say that today we are living on the spiritual capital of our fathers. Our people demand justice, freedom, and respect for the sanctity of personality, but they pay not the slightest heed to the religion which brought these values to the earth and nourished them throughout the ages. They dread the coming of Communism, but all they have to combat it is the faint spiritual hangover of a fast-disappearing tradition.

There are still skeptics who claim that the Christian church rose and grew and spread as a result of a tradition which gathered around the name and memory of Jesus of Nazareth. But this is asking us to believe too much. A secondhand faith would never have inspired men to march out against crushing odds. The secret of success of the early Christians was that the truths they proclaimed were not taken over from others but burned in the very marrow of their bones. "I know whom I have believed," cried Paul, explaining in one brief sentence the impact he made on history. "It is no cunningly devised fable we are giving you," says Peter, "for we were eyewitnesses of Christ's majesty"—again reminding us of the real reason behind the miraculous rise of Christianity. "That which we have heard, which we have seen with our eyes, which our hands have handled of the Word of life declare we unto you," says John, striking the same note of glowing certainty and passionate conviction on which the early Church was built.

There was a time when the Church made history, arrested apathy, and heralded in the dawn of a new age of the spirit, when men arose who could say, "This thing we offer you is not borrowed or secondhand. It can stand the test of crisis, because we have tried it out in the red-hot crucible of personal experience and found it to work." So it was when Columba carried his faith from Ireland to Scottish shores; so it was when St. Francis appeared like a beacon in the Dark Ages of Europe and pointed men to Christ again. So it will be when men and women step forward and strike with their lives, as with their words, the authentic ring of personal conviction. Only then will paganism receive its deathblow and the church of Christ triumph as of yore.

This parable would remind us of:

THE NEED FOR SECRET RESERVES

Both the wise and foolish virgins possessed lamps. Outwardly there was no visible difference between ehem, until the cry at midnight. A lamp burns till the oil that feeds it dries up; then it goes out in flickering spasms. Here we are reminded that what really counts is not the lamp but what feeds the flame.

Outwardly there may seem little or no difference between the decent pagan and the professing Christian. The one appears to be as good, as honest, as reliable as the other, but always it is the midnight hour of crisis and emergency that separates them. Somerset Maugham unwittingly illustrates this in one of his characters. An active, practical, emancipated young woman she was, who spent her energies doing social service in the slums of a city. In time she married and had a child. One night a drunken motorist collided with their car and her husband and baby were killed. The madness of bitterness and despair took possession of her. She sank from one level of moral depravity to another, till at last she committed suicide in the squalid back streets of a southern French town.

Psychologists tell us that we cannot exaggerate the important part the subconscious plays in the formation of character. Nothing is ever forgotten. The thoughts we think, the emotions we experience, the decisions we make, sink down below the threshold of consciousness and, in the hidden depths, they determine our choices and the way we respond to challenge and crisis.

That is why religion has always emphasized the importance of prayer and worship. Their effect is never transient and ephemeral. They help to shape the subconscious side of our nature. They form the oil feeding the lamp of our spirit. They help to build up the secret reserves of personality. Life is never easy, and as we confront its cruel tests we must possess adequate resources to enable us to stand "in the evil day, and having done all, to stand."

CHAPTER TWENTY-THREE

The Heresy of Living in Compartments

"What therefore God hath joined
together, let not man put asunder."
—MARK 10:9

THOSE WORDS WERE FIRST SPOKEN IN ANSWER TO A PARTICU-
lar question put to Jesus by critics who tried to trip him up.
Though they specifically apply to the pressing problem of
marriage and divorce, they cover a much wider range. They
point to the essential oneness of life—a unity which we dissolve
at our own peril.

It is quite incredible that, after two thousand years of Christianity, many of us have still the weirdest notions as to what
the gospel really means. We have been exposed to the Bible from
birth. This book has been in our homes or on our shelves since
we can remember, yet we remain stubbornly blind to the
revolutionary implication of its message. We have been guilty
of the common sin of compartmentalism which splits life up
into watertight sections for our own convenience. We have
forgotten the opening clause of the Apostles' Creed—"I believe
in God, the Father Almighty, Maker of heaven and earth"—
that is to say, creator of matter as well as of mind, author of the
material as well as the spiritual, the architect of the entire structure we call life. We have broken up the fundamental unity of
existence, and history could be defined as the record of that
stupid and impertinent interference. There is, for example, the
divorce between:

SACRED AND SECULAR

Nowhere is this divorce seen more clearly than in the highlands and islands of Scotland. On this point I can speak with

146

a certain amount of authority, for, from early childhood, I have been aware of a sharp line of demarcation drawn between the two. Once a man became a member of the church he gave up such worldly preoccupations as sports and concerts; even the singing of a lovely song like "The Eriskay Love Lilt" was frowned upon. The result is simply disastrous. The majority never join the church at all, and those who do—sincere and saintly though they be—are under pressure to conform to a rigid pattern. The incidence of drink among lusty young men along the West Highland fringe should not surprise us; it is a sort of psychological revenge. If human nature is thwarted in one direction, it inevitably seeks an outlet in another.

And the same benighted heresy of divorcing the sacred and the secular is prevalent in the cultured city of Edinburgh, only it expresses itself in a far more subtle and dangerous way. Supposing you were to take a Gallup poll of the opinions of the church members of Edinburgh as to whether religion and politics should be mixed, I am pretty confident what the answer would be. By and large, the majority would emphatically declare against this alignment. We would be told that our business in the church is to stimulate and develop the spiritual side of human nature, to save men's souls, and proclaim the gospel. The complex questions of wages, foreign policy, social legislation, our attitude toward the Communists, we must leave to the experts, to the politicians whom we pay to look after these things.

This is history's most blasphemous heresy, also its most dangerous. It is diametrically opposed to the teaching of the Bible, which claims that God created everything, that religion and life are synonymous terms, that there is nothing beyond the scope of Providence.

It might not be too extravagant to claim that this divorce is the root cause of all our modern wars. If Christianity in the past had not confined itself simply to the spiritual—if it had boldly staked out God's total claims—if it had interpenetrated the spheres of industry, commerce, and politics in the name of him who is the Author of all things, evil eruptions like Naziism and Communism would not have burst out in our day and age. Christians who oppose mixing up religion with politics are sincere and devout beyond any question, but theologically they are illiterate. It shocks them to hear this preached from a pulpit; but to hear that this was the attitude advocated by Stalin and Hitler in the past, leaves them quite unmoved. Truly nothing can paralyze the mind or blind the eyes of the soul like a stubbornly entrenched prejudice.

We talk of a "universe," not a multi-verse, which means that to God, creator of heaven and earth, all things are one. From there exists an underlying unity between the minutest electron and the mightiest star. Therefore there is no rigid division between body and mind, spirit and matter. Browning, with his Christian insight, expresses this thought in "Rabbi Ben Ezra":

> Let us not always say,
> "Spite of this flesh to-day
> I strove, made head, gained ground upon the
>   whole!"
> As the bird wings and sings,
> Let us cry, "All good things
> Are ours, nor soul helps flesh more, now,
>   than flesh helps soul!"

God created the sacred and the secular and what God hath joined together let no man put asunder.

REASON AND REVELATION

Let us be quite honest: our religion is based on an act of revelation. Julian Huxley has written a book called *Religion Without Revelation,* but that does not make sense to the Christian. We believe that there are definite limits to logic—that the exercise of reason alone can take us so far but no farther; that in the last resort faith is given—it is a flash of revelation from the beyond. When, long before the other, Peter had professed the divinity of Jesus, Christ's answer was very significant: "Blessed art thou, Simon Barjona: for flesh and blood hath not revealed it unto thee, but my Father which is in heaven." But we have forgotten that it is one and the same God who revealed his nature in Christ and discloses himself in the operations of human reason. Reason and revelation are both of God. From the beginning they were joined together, but man has often insisted in putting them asunder.

Theology was once regarded as the "Queen of the Sciences." Now it has come to be looked on as the Cinderella, and the blame for this shameful relegation the Church herself in large measure must accept. Through the centuries, the Church passionately opposed many of the major efforts made by man in the quest for truth. She opposed Galileo and Copernicus in their new conception of the universe, Darwin in his revolutionary theory of evolution, Robertson of Aberdeen in his historical approach to the Bible. This fight to the death against every advance of knowledge sprang out of fear—fear that reason, if given free play, would destroy religion and discredit revelation. At bottom it is essentially a lack of faith.

I profoundly believe that theology is still queen of the sciences, not merely in name but in fact. Start from any angle you like, and if you pursue the question asked to its logical conclusion, you find yourself in the realm of theology. The psychologist talks of integration of personality as the ideal, but when he asks himself the question "Why integration?" he has crossed the boundaries of psychology into theology. The doctor assumes health of body and mind is the "norm," but when he asks why it should be the norm, he has ceased to be a doctor and has become a theologian in the making. The nuclear physicist tells us that the atom behaves according to certain laws, but when he asks why it should respond to any law he is tacitly assuming that theology is still queen of the sciences. Every question—be it social, scientific, political, or economic— now becomes ultimately a theological one.

More than once I have been asked recently if I am in agreement with certain famous evangelical campaigns. My answer is that I support any evangelical campaign that does justice to the two dimensions of God's own nature—revelation and reason. We who preach must be evangelical in the profoundest sense of that term; we must be heralds of the good news—what the New Testament means by the kerugma. At the same time, like these superb evangelists Pascal and Henry Drummond, we must address ourselves to people whom God has endowed with rational faculties. Any school of evangelism that belittles or despises the reason which has invented television and discovered how to split the atom is not worth listening to; it is, in fact, already under a sentence of death. We who believe must never be on the defensive. We know that behind our own partial, fragmentary knowledge exists one Absolute Mind, that reason and revelation are both of God. "What . . . God hath joined together, let not man put asunder."

COMFORT AND CHALLENGE

The two are inextricably mixed up. Examine every cardinal Christian doctrine which is a source of comfort, and, implicit in it, is an inescapable challenge. Take the central Christian doctrine from which all the others stem—that of the fatherhood of God. How immeasurably comforting to know that ultimately we are at the mercy not of blind chance but of a Father who pities us even as an ordinary father pities and protects his children. Without this basic belief there can be no comfort. But think of the exacting challenge! If we really believed it, we would not send some of God's children to good schools and others to inferior ones where the handicap is higher. If we really believed it, we would not sleep at night knowing that millions of our brothers and sisters go from one end of the year to another without a square meal. If we really believed it, we would rise in arms against the sins of privilege, exploitation, and social injustices. If we really believed it, we would not talk so

self-consciously of foreign missions—nor would we tolerate
the opinion of those glib people found in every church who do
not see the point of spending money on the heathen abroad
while there is so much to do at home. Nobody has a right to
such opinions any more than they have to the opinion that
the earth is flat. Paradoxical, isn't it, that the most comforting
of all beliefs in actual fact is also the most challenging?

Again, there is the doctrine of immortality. Let us get the
theology straight here. It doesn't stand on its own legs: it is
a corollary from a more primary belief, the fatherhood of God.
But what a source of strength and comfort it can be to those
broken by bereavement, the most cruel and shattering of all
blows. Discredit it, and life loses its point. The poet is right:

> Sad as the wind that blows across the ocean,
> Telling the earth the sorrow of the sea.
> Vain is my strife, just idle empty motion,
> All that hath been is all there is to be.

What a difference it makes to believe that those whom "we
have loved long since and lost awhile" have not been blotted
out, but live the fuller life—where one day we will experience
with them what the Creed calls "the communion of saints."

But think for one moment of the towering challenge of such
a belief. This life is but the preliminary training-ground for
the next. One day we shall meet not only our loved ones but
also God, and, in his presence, how shall we stand? Karl Marx
and his satellites are simply romancing; the doctrine of belief
in immortality, properly understood, heightens, not diminishes,
our sense of moral responsibility. It means that we live every
moment of our lives in Augustine's familiar phrase *"sub specie
aeternitales."* If our generation has lost all sense of moral responsibility—as all the pessimists aver—it will only return when
all people have recovered what the New Testament means by
faith in the life everlasting. Let me put it quite boldly and

bluntly. You cannot live as you ought, you cannot do justice to the range and sweep of personality, you cannot feel morally responsible in any profound or ultimate sense till you come to terms with immortality. This life and the next are one, and what God hath joined together let no man put asunder.

CHAPTER TWENTY-FOUR

What Is Your View of Religion?

"I am the bread of life."
—John 6:35

THERE USED TO FLOURISH A SCHOOL OF THOUGHT WHICH depicted Jesus as the most liberal and reasonable of men. Far from being extremist or extravagant, he was in all things moderate, the ardent exponent of the middle way. No wonder, then, so many eloquent sermons were preached on the reasonableness of Christ's claims and the common-sense wisdom of Christianity.

The New Testament, however, under close scrutiny does not support such a contention. Wisdom and balance and harmony certainly there are, but they do not obscure for us one challenging fact which it is impossible to side-step. I mean the extraordinary self-consciousness of Jesus.

Shakespeare passed on to posterity Macbeth, Hamlet, Othello, King Lear, and a gallery of characters whom we know more intimately than we know our friends: but of the man himself we know next to nothing. Would-be biographers are still busy gathering scattered scraps of fact and fiction in a desperate bid to reconstruct a personal history. It looks as if the poet deliberately drew a curtain of anonymity over his own personality, leaving future generations to guess and speculate.

Christ's emphasis was entirely different. He, too, was an artist—the perfect storyteller with an uncanny sense of the dramatic and a wonderful mastery of language, yet he spent his energy not in composing poetry or writing books but in impressing his personality on his world and generation.

Jesus, unlike Shakespeare, will not remain anonymous. He is not prepared to share the honors with anyone else. He leaves no room for confusion. In a series of staggering claims, he makes it perfectly plain that he regards himself as standing in a special relationship with God. "Before Abraham was, I am." "He that hath seen me hath seen the Father." "I am the light of the world." "I am the bread of life."

And the question is, Can we take him seriously? Is George Bernard Shaw right in hinting that, up to the incident at Caesarea Philippi, Jesus was pre-eminently sane, but after Peter's affirmation—"Thou art the Christ, the Son of the living God"—he became obsessed with a conviction of his own divinity and thereafter talked of it continually? Is this the truth of the matter? Are all our theological speculations and time-honored creeds mere examples of misguided effort? Must we pass this Jesus off as a monomaniac who has deceived the succeeding generations and still appeals to some subintellectual instinct in man? Let us, then, take some of the contemporary views of religion and hold them up against Christ's own conception of his stature and place among men.

There is the view which might be described as:

Religion as Fantasy

This theory has peculiar attraction for the modern mind. As a religion of realism it claims disciples by the million. Some of the most eloquent prophets of our age proclaim it as the only liberating gospel. "Son of man," they cry, "stand upon your own feet and shake off this shackling superstition: then, and then only, can you be saved."

Sponsoring it are two well-known and powerful schools of thought. Karl Marx and his followers declare that religion has deliberately deluded men by offering them the bliss of a problematical world to come in place of the equity and higher material comfort which should be their lot here and now. Even more devasting is the attack made by Freud. Psychology freely admits that religion helps by offering us comfort, encouragement, and a measure of serenity in a hard and cruel world; but it has this effect not because it is objectively true but because subjectively it acts as a stimulant. It is a "shot in the arm"—a process of autosuggestion. Jung, the famous psychiatrist, has confessed in one of his books that, although he himself cannot subscribe to the Christian creed, he always advises his patients to take up religion, because as a doctor he had discovered that religion was "a useful illusion."

If I may be permitted, I will illustrate what this means from my own experience. During the war, while training as a paratrooper, I became friendly with a man who was terrified of jumping from an airplane. He could not sleep at night, he could not eat, and one saw him degenerate into a bundle of nerves. Morbidly he would burst into a conversation. "I know when the test comes I'll refuse to jump," he would cry out hysterically. The morning we made our first drop he went up to the jump-master. "Sergeant," he said, "promise me that if I refuse to jump you'll give me a push." Always an accommodating fellow, the sergeant replied, "With the greatest of pleasure." I can still see him sitting opposite me across the hole in a Whitley bomber through which we had to go, his knuckles standing out white as tensely he gripped the edge, the pupils of his eyes dilated with fear. The light changed from red to green. Above the roar of the propellers came the order—"Go!" He sat still paralyzed with fright. The sergeant gave him a push and he disappeared from sight. A few minutes later, as we sat in the canteen drinking tea, he went up to the sergeant. "Thank you for the push, Sergeant," he said. Round-eyed with surprise, the sergeant looked at him. "What push?" he asked innocently. From that moment this man became one of the best paratroopers in the battalion.

A useful illusion. And, so say the critics, religion is like that. It is useful to believe that a God exists somewhere behind the perplexing panorama of our existence. In the strain and stress of living it is useful to feel that there are everlasting arms underneath you bearing you up and that there is a haven of refuge into which our tempest-tossed souls creep when the end comes. But it is all sheer fantasy. If this view is correct, all we can say is that the sanest and strongest personalities in history were cruelly deceived—that there is a radical twist at the very heart of our existence—that Shakespeare's Macbeth was right: Life is "a tale told by an idiot, full of sound and fury, signifying nothing."

There is another view which sees:

RELIGION AS A LUXURY

Of the two, the equating of religion with luxury is by far the more devastating. The sociologists and psychologists may loudly talk of fantasy, but if this is not the truth, they cannot in the end do much damage. But this popular cult of regarding Christianity as an extra and unimportant addendum—a mere embellishment—is in danger of corroding the foundations of our faith and reducing the cross of Christ to an absurdity.

Unfortunately, it is an attitude which meets us on every conceivable level of life. It does not attack or denounce. Why should it? It possesses a far more deadly weapon—a smooth, well-bred, casual indifference. If it deigns to argue at all, it does so on these lines: Some people can't live what they call the full life without cultivating art, literature, philosophy, and classical music, but the average man going about his daily business can live a robust, full-blooded life without ever troubling himself with these pursuits. Similarly, some people feel that you cannot do justice to the full range of personality without practicing some form of religion. They constitute but an eccentric few, and toward them, as to all eccentrics, we must cultivate the grace of tolerance.

This view of religion is not the sole prerogative of the pagan masses and clever intellectuals. It is the unconscious attitude of most church members in Britain and expresses itself in a multiplicity of gestures.

There is the crisis in the number of candidates available for the ministry in every single denomination in the country. To be sure, social and economic factors enter into the situation; but the problem remains, primarily, a spiritual one. Behind the dearth of ministers and the poor church attendance in Scotland lies an attitude of mind which looks at religion not as a matter of life and death but as a luxury which can be dropped if other things intrude.

How can the average church member square his conscience over the amount he spends on smoking and pleasure compared to the amount he spends on the church of Christ in this godless world? If Communism ever conquers, trampling our values underfoot, the explanation will be simple: professing Christians have not looked on faith as a matter of life and death. They have taken the religion that made Jesus sweat blood in Gethsemane and emptied it of means and power by using it intermittently as a luxury they can dispense with, at will.

Religion as a Basic Necessity

This surely is what Jesus meant when he used the bold and brilliant metaphor, "I am the bread of life." Personally, I never grasped the full meaning of this breath-taking claim till I became a prisoner of war in Germany during the last war and saw things happening which, in normal days, I would have said were incredible. I have seen civilized and cultured men fighting over a morsel of bread. I have seen graduates of famous universities gambling with cards to see who would get the end slice of a loaf. For three months on end I could not sleep at night for hunger, and when I did drop into an uneasy sleep I dreamed not of sweets and cakes and pleasant tidbits, but of an honest-to-goodness chunk of bread I could crunch between my teeth.

No one who has lived on the raw edge of life can ever fail to realize that bread is not a luxury but a basic necessity.

"I am the bread of life," says Jesus. "Without me the soul starves, the spirit of man shrivels up, and personality is forever dissatisfied." It is a colossal claim. But is it true? Is it just metaphor and nothing more? This Western civilization of ours, cringing under the shadow of the atom, is near the brink of destruction. What do you think will save it? Not science or technical superiority. It needs Christ. This land of ours which we love best, where crime and gambling and irresponsibility are on the increase—what does it need most? Not more education and culture. It needs Christ. I don't know about you; you may be able to meet "the slings and arrows of outrageous fortune" and stem the torrential floods in your own strength, but I cannot. To remain true to my vows, to keep my vision and courage and love for men, to do justice to all the latent possibilities within, I need the stamina of Christ. And I suggest that you do too, in success and failure, in joy and sorrow, in life and death, you need him who is the basic stuff of our existence —the Christ who said, "I am the bread of life."